International Conference on Bio-Medical Instrumentation and Related Engineering and Physical Sciences 2015 (BIOMEP 2015)

Journal of Physics: Conference Series Volume 637

Athens, Greece
18 - 20 June 2015

ISBN: 978-1-5108-1315-1
ISSN: 1742-6588

Printed from e-media with permission by:

Curran Associates, Inc.
57 Morehouse Lane
Red Hook, NY 12571

Printed by Curran Associates, Inc. (2016)

Institute of Physics
Dirac House, Temple Back
Bristol BS1 6BE UK

Phone: 44 1 17 929 7481
Fax: 44 1 17 920 0979

techtracking@iop.org

Additional copies of this publication are available from:

Curran Associates, Inc.
57 Morehouse Lane
Red Hook, NY 12571 USA
Phone: 845-758-0400
Fax: 845-758-2634
Email: curran@proceedings.com
Web: www.proceedings.com

Table of contents

Volume 637

International Conference on Bio-Medical Instrumentation and related Engineering and Physical Sciences (BIOMEP 2015)
18–20 June 2015, Athens, Greece

Accepted papers received: 31 August 2015
Published online: 16 September 2015

Preface

Papers

Journal of Physics: Conference Series **637** (2015) 011001 doi:10.1088/1742-6596/637/1/011001

Preface/introduction

The International Conference on Bio-Medical Instrumentation and related Engineering and Physical Sciences (BIOMEP 2015) took place in the Technological Educational Institute (TEI) of Athens, Greece on June 18–20, 2015 and was organized by the Department of Biomedical Engineering.

The scope of the conference was to provide a forum on the latest developments in Biomedical Instrumentation and related principles of Physical and Engineering sciences. Scientists and engineers from academic, industrial and health disciplines were invited to participate in the Conference and to contribute both in the promotion and dissemination of the scientific knowledge.

Editors' short CVs

Professor Ioannis Kandarakis: Current position: Professor at the Technological Educational Institution (TEI) of Athens, Department of Biomedical Engineering (http://www.bme.teiath.gr) / Laboratory of Radiation Physics, Materials Technology and Biomedical Imaging). Educational qualifications: Physics and Medical Radiation Physics, postgraduate and doctoral degree (University of Patras and Université "Paul Sabatier" – Toulouse III). Research and Scientific Activities: 1. Medical Imaging Detectors: scintillator and phosphor materials evaluation for application in x ray and gamma ray detectors, 2. Evaluation of medical imaging systems using objective image quality metrics, 3. Non-ionizing radiation applications (Magnetic Resonance Imaging and Ultrasonic Imaging). Scientific Publications: 110 original papers in peer reviewed international scientific journals, more than 100 conference proceedings and abstract books, 460 citations to published work (excluding self-citations). 14 research projects. Greek language textbooks (Physics of Diagnostic Radiology, Physics of Nuclear Medicine, Physics of MRI). Reviewer in 18 scientific journals. Teaching experience at both undergraduate and postgraduate level in Medical Radiation Physics, General Physics and Medical Imaging Systems. Supervision of 14 doctorate theses. Member of conference committees, session chairman and invited speaker in more than 20 international conferences. Working experience in Medical Radiation Physics departments in Greek University Hospitals.

Journal of Physics: Conference Series **637** (2015) 011001 doi:10.1088/1742-6596/637/1/011001

Dr. Ioannis Valais: Current position: Associate Professor at the Technological Educational Institution (TEI) of Athens, Department of Biomedical Engineering. Educational qualifications: Graduate (1991) of Department of Medical Instruments, Technology School of Applied Technology, TEI of Athens, holds a Master's Degree (Masters) of Science in Biomedical Engineering and Medical Physics from the University of Aberdeen, UK, 1993 and a Ph.D. in the Department of Medical Physics, School of Medicine, University of Patras, 2008. From 1993 to 2001 worked in the technical support and service division of a company trading imaging and therapeutic medical equipment. After many installations of, and training in the technical support for Medical Equipment, he arrived in 1998 to the post of Head of Technical Support Systems in Nuclear Medicine, Lasers and Peritoneal Dialysis. From 1995 to 2000 he worked as part time lecturer in the Department of Medical Instruments Technology, TEI of Athens, where he taught laboratory courses Service and Readiness of Medical Equipment and Constructions.

From 2001 to September 2009 worked as a full time Lecturer in the Department of Medical Instruments Technology, TEI of Athens, teaching laboratory courses Constructions I & II and Readiness and Maintenance of Medical Equipment. From October 2009 to November 2013 worked as an Assistant Professor in the Department of Medical Instruments Technology, TEI Athens, teaching theory and lab courses Maintenance Quality Control of Biomedical Equipment and Methodology in the Design and Construction of Biomedical Devices. Since November 2013 works as an Associate Professor in the Department of Biomedical Engineering T.E. (former Medical Instruments Technology), TEI Athens, teaching theory and lab courses on Maintenance & Quality Control of Biomedical Equipment, Optoelectronics and Medical Laser, and Methodology in the Design and Construction of Biomedical Devices.

Areas of research: Construction and evaluation (both experimental and theoretical in terms of absolute efficiency, MTF, NNPS, DQE & IC) of scintillating screens for use in digital imaging systems, image quality evaluation in PET & SPECT systems, Monte Carlo methods, Dual Energy imaging, electronic and mechanical engineering on the improvement of medical equipment, quality assurance for nuclear medicine imaging systems, design protocols and maintenance of quantitative and qualitative checks in medical equipment. Collaboration with other departments in the TEI of Athens on new imaging and diagnostic applications. Supervision and seeking graduate work on a related theme is a part of research interests. Funded research projects: 15 Greek projects as a researcher and two programs approved at the first stage of evaluation.

Scientific Publications: he has published 52 peer reviewed papers.

http://www.teiath.gr/stef/tio/staff/en_CV_Valais_Ioannis.pdf

Journal of Physics: Conference Series **637** (2015) 011001 doi:10.1088/1742-6596/637/1/011001

Dr. George Fountos was born in 1965. He is an Associate Professor in the Department of Biomedical Engineering, Technological Educational Institute of Athens, since 2007. He received his Diploma in Physics from the University of Ioannina, Greece, in 1987, his M.Sc. in Medical Physics in 1989 from Patras University and his Ph.D. in Medical Physics in 1995 from the University of Ioannina. He received his licence in Radiation Physics in 1996 and he is the Treasurer of the Hellenic Association of Medical Physics (HAMP) from 2004. He has published 59 full papers in international journals and books and has many publications in conference proceedings (265 citations in published work). His research interests are focused on Dual Energy Body composition, quality assurance and simulations in ionizing Radiation Medical Systems, radiation protection, physics of nuclear medicine and radiology, electron microscopy (TEM) etc.
e-mail: gfoun@teiath.gr

Dr. Michail Christos: Current position: collaborating researcher at the Technological Educational Institution (TEI) of Athens, Department of Biomedical Engineering. Educational qualifications: first degree in Biomedical Engineering (Athens, Greece), Master's degree and PhD in Medical Physics. Awards: He was awarded with a scholarship from the Greek State Scholarships Foundation (I.K.Y.) for the academic years 2007–2010 for the 'Technology of Telecommunications in Medicine' specialty after succeeding in the Foundation's exams. Areas of research: construction and evaluation (both experimental and theoretical in terms of absolute efficiency, MTF, NNPS, DQE & IC) of scintillating screens for use in digital imaging systems, image quality evaluation in PET & SPECT systems, Monte Carlo methods, Dual Energy imaging. Funded research projects: 13 Greek projects as a researcher and two programs approved at the first stage of evaluation. He has also participated as a reviewer for scientific journals (European Radiology, Measurement, the Journal of Engineering Research, Physica Medica, Ecotoxicology and Environmental Safety, Biomed Research International & NIMA) and proposal reviewer. He is also a member of the

Journal of Physics: Conference Series **637** (2015) 011001 doi:10.1088/1742-6596/637/1/011001

Editorial Board of the scientific journal Engineering, Technology and Applied Science Research (ETASR).
Scientific Publications: He has published 44 peer reviewed papers. Also he has 18 publications in international scientific journals without impact factors. Furthermore 57 proceedings in international scientific conferences with referees, 59 abstracts, eight book chapters, 16 proceedings in national scientific congresses and nine invited lectures. The number of citations given in his papers, after a relatively short time of publication, is 179 (excluding self-citations) and 404 in total.
e-mail: michail@upatras.gr
https://scholar.google.gr/citations?user=lCQiC6QAAAAJ&hl=el&oi=ao
http://www.teiath.gr/stef/tio/IonRadWebSite/en/tstaff/cv_michail.pdf

Part3: Clinical potential
UCL

Objectives (specific)
• To objectively define the voice pleasantness concept
– for European Portuguese and
– multilanguage perspective;
• To classify and provide intensity estimates for a given voice in terms of its perceptive pleasantness;
• To understand voice pleasantness correlations between
– prosodic parameters and
– acoustic features;
• To understand the potential suitability of a voice for a TTS system;
• To study the application of speech enhancement strategies in TTS systems using the voice pleasantness knowledge;

Journal of Physics: Conference Series **637** (2015) 011001 doi:10.1088/1742-6596/637/1/011001

Sponsor or funding acknowledgements

The conference was founded by the European Union (European Social Fund) and Greek national resources under the framework of the "ARISTEIA" project MISCIRLU code 1476 of the "Education & Lifelong Learning" Operational Program.

The conference organizers also express their gratitude to the Education and Research Committee of the Technological Educational Institute (TEI) of Athens for its contribution in publishing the present volume

Lists of committees/participants

BIOMEP 2015 - Committees
Conference Chair:

Prof. I. Kandarakis (Chair), TEI of Athens, Greece
Prof. E. Ventouras (co-Chair), TEI of Athens, Greece

Organizing Committee:

Assoc. Prof. I. Valais (Chair), TEI of Athens, Greece
Asst. Prof. I. Kalatzis, TEI of Athens, Greece
Asst. Prof. N. Kalyvas, TEI of Athens, Greece
Lect. P. Liaparinos, TEI of Athens, Greece
Dr. E. David, TEI of Athens, Greece
Dr. Ch. Michail, TEI of Athens, Greece

Scientific Committee:

Assoc. Prof. G. Fountos (Chair), TEI of Athens, Greece
Prof. I. Damilakis, University of Heraklion, Greece
Dr. A. Gektin, Institute for Scintillation Materials of National Academy of Science of Ukraine, Ukraine
Prof. P. Giannakopoulos, TEI of Piraeus, Greece
Prof. G. Kaltsas, TEI of Athens, Greece
Prof. C. Kappas, University of Thessaly, Greece
Prof. K. Kourkoutas, TEI of Athens, Greece
Prof. E. Patsavoudi, TEI of Athens, Greece
Prof. I. Sianoudis, TEI of Athens, Greece
Prof. R. Speller, University College London, UK
Prof. E. Zych, University of Wroclaw, Poland

Honorary Committee:

Prof. D. Triantis, TEI of Athens, Greece
Prof. D. Vattis, TEI of Athens, Greece

Journal of Physics: Conference Series **637** (2015) 011002 doi:10.1088/1742-6596/637/1/011002

Peer review statement

All papers published in this volume of *Journal of Physics: Conference Series* have been peer reviewed through processes administered by the proceedings Editors. Reviews were conducted by expert referees to the professional and scientific standards expected of a proceedings journal published by IOP Publishing.

Journal of Physics: Conference Series **637** (2015) 012001 doi:10.1088/1742-6596/637/1/012001

Biomedical Impact in Implantable Devices-The Transcatheter Aortic Valve as an example

Alexandros Anastasiou[1], George Saatsakis[2]

[1]Technological Educational Institution of Athens (Agiou Spyridonos 57, 12243, Athens), Biomedical Engineering Department, Greece
[2]University of Athens, Medical School, Aretaieio University Hospital (Vasilis Sofias Ave. 67, 11528, Athens), Biomedical Engineering
Department, Greece

E- mail: alexandros_anastasiou@outlook.com

Objective: To update of the scientific community about the biomedical engineering involvement in the implantable devices chain. Moreover the transcatheter Aortic Valve (TAV) replacement, in the field of cardiac surgery, will be analyzed as an example of contemporary implantable technology.

Methods: A detailed literature review regarding biomedical engineers participating in the implantable medical product chain, starting from the design of the product till the final implantation technique.

Results: The scientific role of biomedical engineers has clearly been established. Certain parts of the product chain are implemented almost exclusively by experienced biomedical engineers such as the transcatheter aortic valve device. The successful professional should have a multidisciplinary knowledge, including medicine, in order to pursue the challenges for such intuitive technology. This clearly indicates that biomedical engineers are among the most appropriate scientists to accomplish such tasks.

Conclusions: The biomedical engineering involvement in medical implantable devices has been widely accepted by the scientific community, worldwide. Its important contribution, starting from the design and extended to the development, clinical trials, scientific support, education of other scientists (surgeons, cardiologists, technicians etc.), and even to sales, makes biomedical engineers a valuable player in the scientific arena. Notably, the sector of implantable devices is constantly raising, as emerging technologies continuously set up new targets.

1. Introduction

Each year millions of patients improve their quality of life through surgical procedures that involve implanted medical devices. The term implant is used for devices that replace or act as a fraction of the whole biological structure. Currently, implants are being used in many different parts of the body for various applications such as orthopedics, pacemakers, cardiovascular stents, defibrillators or neural prosthetics. While they can save or improve the quality of a patient's life, they also pose some of the biggest engineering challenges which include design, prototyping, selection of materials, manufacturing and testing [1]. The biomedical engineering involvement in medical implantable devices is the most appropriate in such tasks starting from the design and extending to the development, clinical trials, scientific support, education of other scientists (surgeons, cardiologists, technicians etc.), and even to sales. This conclusion makes biomedical engineers a valuable player in the scientific arena of implantable devices.

2. Methods

The contribution of biomedical engineer in the field of implantable devices is valuable, starting from the idea capturing to the education of other scientists for the proper positioning of the device. In this review will be discussed life-threatening diseases, in the field of contemporary cardiac surgery, such as aortic stenosis. The main focus of this paper is to analyze the biomedical impact throughout the product chain. Furthermore, essential features related to transcatheter heart valve such as hemodynamics and clinical trials will be discussed.

3. Heart Valves

The aortic valve is one of the four valves in the human heart and through which blood flows from the heart to the whole body. Aortic stenosis is the condition whereby the main valve leading from the heart to the rest of the body becomes tight, and it's hard for the heart to eject blood into the circulation system for the rest of the body. At this stage aortic stenosis is generally caused by the calcification of the valve leaflets and with the deposition of the calcification the leaflets become immobilized. The symptoms related to aortic stenosis are often shortness of breath, chest pain, swelling of the legs, even syncope. Medical treatment for severe symptomatic aortic stenosis is not effective, and without aortic valve replacement, the rate of mortality is approximately 25 percent in 1 year and 50 percent in 2 years. Surgical replacement of the aortic valve markedly improves the symptoms of the patient and the patient's long survival [2]. In normal patients, these operations carry a perioperative mortality of about 3 to 4 percent. However, according to recently published data of the European Society of Cardiology, 33 percent of all adults ≥75 years old with severe aortic stenosis do not undergo surgery even though it is the only successful treatment. The open surgery entails substantial risks for some patients with severe comorbidities such as advanced age, patients with previous history of open heart surgery etc., and for some considered as "high risk" patients [3]. To meet the medical needs of this population, a new technology has emerged over the past decade and is now being put into clinical practice: transcatheter aortic valve replacement (TAVR). This technology has significant impacts throughout the health care field with the creation of a new biotechnology industry around transcatheter valves, the creation of multidisciplinary "heart teams", and the construction of hybrid procedure rooms with both cath lab and operating room capabilities.

4. Design of Transcatheter Aortic Valves

Both the Edwards SAPIEN valve and the Medtronic CoreValve are designed to function through a mechanism similar to a normally functioning human tricuspid aortic valve. However, while both are trileaftlet in design with a metallic framework for support, their construction as well as preparation and delivery have significant differences. The integrated Edwards SAPIEN XT transcatheter heart valve system is comprised of bovine pericardium which is individually matched for flexibility and strength. The valve tissue is affixed to a stainless steel frame with cobalt-chromium alloy. This frame design allows the valve to be compressed down further than the older generation of Edwards SAPIEN valve during delivery and catheter implantation (the first generation used stainless steel frame), so the procedure can be performed in patients with smaller femoral arteries [4]. In contrast to the Sapien valve, the CoreValve is a self-expanding valve with a nitinol frame. Although the first generation of the CoreValve (first implanted in humans in 2004) was made from bovine pericardium with an intra-annular valve function similar to that of the Sapien valve, the current generation CoreValve is made from porcine pericardium. The choice of nitinol as opposed to stainless steel gives the CoreValve system the ability to be loaded onto a catheter delivery system that does not require a balloon and enables the valve to be gradually deployed in stages.

5. Delivery System

Prior to delivery, the Edwards SAPIEN XT valve is tightly compressed using a crimping mechanism onto a balloon catheter. A catheter is a long, thin, flexible tube that is able to guide the new valve through the artery and place it in the correct position. Furthermore, aortic balloon valvuloplasty is performed, and thereafter the Sapien transcatheter heart valve system is advanced up the aorta and placed across the native aortic valve. The balloon with the attached valve is then rapidly inflated and deflated, expanding and releasing the Sapien valve. The newly functioning valve is passively secured to the underlying native leaflets and to the aortic annulus as a result of this delivery. In contrast to the SAPIEN valve, the CoreValve is deployable via self-expansion without the aid of a balloon catheter. The stent is initially crimped and placed in a sheath before being deployed at the treatment site. This is possible due to nitinol's shape memory property that allows the stent to be expanded to its predetermined shape by allowing blood to come into contact with it. Neither of the techniques requires the removal of the native heart but is pushed to the walls of the heart by the artificial valve. The catheter can be inserted into the body from one of several locations such as transfemoral, transapical and transaortic approach.

6. Clinical Trials with TAVR

The safety and effectiveness of the implantable valves has been studied in more than 50,000 patients. The PARTNER Trial was enrolled between 2007 and 2009 and the results were dramatically positive of the transcatheter valve. Up to date, the PATNER II trial randomized 560 patients who were eligible for surgery to transfemoral approach using the new Sapien XT or transfemoral using the original Sapien device. Mean age of the patients was 84 years old and they followed for a minimum of 1 year. Data at 30 days revealed that mortality (Sapien 5.1%, Sapien XT 3.5 %) and appearance stroke (Sapien 3% and Sapien XT 3.2%) were similar between groups. The primary end point in 1 year did not differ between groups (Sapien 34.7%, Sapien XT 33.9%).

7. Results

The scientific role of biomedical engineers has clearly been well established. Since the day of the first pacemakers, implantable devices have undergone a major transformation. Certain parts of the product chain are implemented almost exclusively by experienced biomedical engineers such as the transcatheter aortic device. The successful professional should have prior experience with transcatheter heart valves which includes extensive knowledge of the various access routes as well as anatomy and physiology of the cardiovascular system. To obtain the best outcome for the patient, there must be a true collaboration between surgeons, cardiologists, technicians etc., not only in valuation of the patient but in the procedure itself. Transcatheter aortic valve implantation represents a promising, new technology that aims to replace the high risk surgical heart valve replacement. However, there remain significant difficulties to be overcome. One of the major drawbacks is the issue of paravalvular leaks which is possible due to the valve not being in the correct position. Valve design may also have a positive contribution in resolving this issue. Paravalvular leakage greatly reduces the efficiency of the valve. Another issue that should be examined is the incidence of a stroke. Stroke is a major concern in transcatheter aortic valves as its incidence is higher in patients receiving this method than those surgically or medically treated as observed in the PARTNER study. Another issue is the increased incidence in permanent pacemaker placement for patients getting the CoreValve, as the ability to reposition if needed as well as issues of alternative access in patients whose femoral anatomy cannot accommodate the large sizes required for the transfemoral approach. Lastly, as modifications and improvements are made to the design, the medical field must continue to adapt and develop in response to this technology [5].

8. Conclusions

One of the fastest growing fields of technology, a field of astounding recent achievements and even more ambitious hopes is the biomedical engineering. In rapidly growing field of science, the successful professional should have a multidisciplinary knowledge in biological and health sciences as well as the ability to communicate with life sciences professionals, physicians in order to pursue the challenges for such intuitive technology. Biomedical engineers are having a key role in the chain of implantable devices in general, and more specifically, the transcatheter aortic valves. To fulfill the project they need to cooperate with cardiologists, cardiac surgeons, anesthesiologists, image specialists, nurses and technicians. As the sector of implantable devices expands, the biomedical engineering role will be greater and most valuable in the scientific arena.

9. References

[1] Regar E, Sianos G and Serruys P W 2001 Br. Med. Bull. **59** 227.

[2] Colytaux RR, Williams Jr. JW, Gray RN, Wang 2010 Annals of Internal Medicine. **153** 314.

[3] Lung B, Cachier A, Baron G, Messika - Zeitoun D, Delahaye F, Tornos P, Gohlke – Barwolf C, Boersma E, Ravaud P, Vahanian A 2005 European Heart Journal. **26** 2714.

[4] K Spargias, A Halapas, M Chrissoheris, N Bouboulis, S Skardoutsos, J Nikolaou, A Tsolakis, C Mourmouris, S Pattakos 2014 Hellenic J Cardiol.

[5] Holmes DR Jr. 2012 J Am Coll Cardiol. **59** 1200.

Integrated system for remotely monitoring critical physiological parameters

S Alexakis[1], S Karalis[1] and P Asvestas[1]
[1] Department of Biomedical Engineering,
Technological Educational Institute (TEI) of Athens, Agioy Spyridonos str, 12243 Egaleo, Greece

Email: sotial77@gmail.com, s_karalis19@yahoo.gr, pasv@teiath.gr

Abstract. Monitoring several human parameters (temperature, heart rate, blood pressure etc.) is an essential task in health care in hospitals as well as in home care. This paper presents the design and implementation of an integrated, embedded system that includes an electrocardiograph of nine leads and two channels, a digital thermometer for measuring the body temperature and a power supply. The system provides networking capabilities (wired or wireless) and is accessible by means of a web interface that allows the user to select the leads, as well as to review the values of heart rate (beats per minute) and body temperature. Furthermore, there is the option of saving all the data in a Micro SD memory card or in a Google Spreadsheet.

The necessary analog circuits for signal conditioning (amplification and filtering) were manufactured on printed circuit boards (PCB). The system was built around Arduino Yun, which is a platform that contains a microcontroller and a microprocessor running a special LINUX distribution. Furthermore, the Arduino Yun provides the necessary network connectivity capabilities by means of the integrated Wi-Fi and Ethernet interfaces. The web interface was developed using HTML pages with JavaScript support.

The system was tested on simulated data as well as real data, providing satisfactory accuracy regarding the measurement of the heart rate (±3 bpm error) and the temperature (±0.3°C error).

1. Introduction

Monitoring of vital signs (heart rate, respiratory rate, blood pressure and body temperature) allows assess the general physical condition of a person, provides indications to possible diseases, and helps evaluating the progress toward recovery. The monitoring is accomplished by means of specialized devices, known as medical monitors, which consist of sensors, processing units and display components [1], [2]. Furthermore, medical monitors can be supported by communication links that enable displaying or recording the results through the local network of the hospital.

This paper presents the design and implementation of a custom made portable monitoring device that includes an electrocardiograph of nine leads and two channels, a digital thermometer for measuring the body temperature and a power supply. The system integrates networking capabilities and is accessible remotely allowing the user to select leads, as well as to review the values of heart rate and body temperature.

There are similar projects that implement vital signals monitoring devices, supporting wireless transmission [3], [4]. Comparing to these projects, the proposed implementation provides improvements with respect to storing and transmitting the data. More specifically, the server and monitor are in the same device and multiple storing methods are supported. Furthermore, it is provided the capability to select ECG leads remotely.

2. Materials and Methods

The architecture of the proposed device is shown in figure 1. There is a battery regulated voltage supply of ±5V, minimizing the possibility of electric shock and making the device completely portable and functional for at least 5 days. The voltage supply is protected, by two diodes, from improper battery installation and there is a red LED that is activated by the microcontroller in case of low battery and the indication in case of battery replacement need.

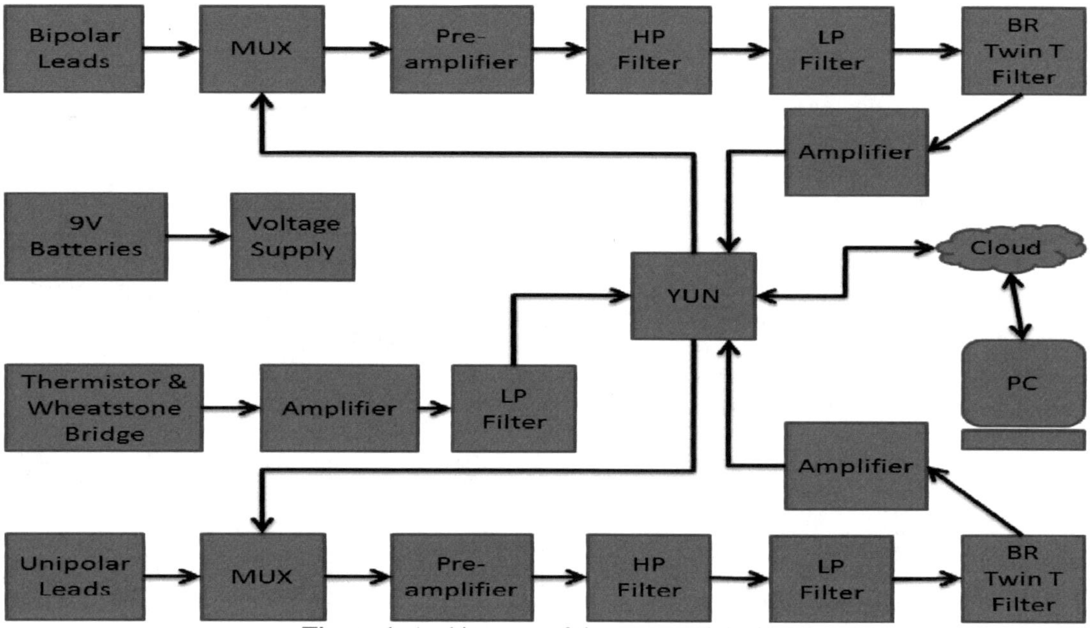

Figure 1. Architecture of the proposed device.

The digital thermometer is designed to operate in the range of 25°C to 43°C and is implemented by means of a thermistor and a Wheatstone bridge. The output of the Wheatstone bridge is amplified by an instrumentation amplifier and then is filtered by a second order low-pass (LP) Butterworth filter, with cutoff frequency at 20Hz, in order to clean the signal from high-frequency noise.

The thermometer was calibrated by means of an Hg analog thermometer immersed in a tank of deionized water. A special device was used to maintain the temperature of the water to a specific value, selected by the user. The voltage output of the digital thermometer was measured for specific reference values in the range of interest. The equation, through the graph voltage – temperature, is used by the microcontroller to convert the measuring voltage into temperature.

The electrocardiograph supports the acquisition of ECG signals from bipolar or unipolar leads. The bipolar leads are connected to a multiplexer (MUX) with eight inputs and two outputs, whereas the unipolar leads are connected to a MUX with eight inputs and one output. The multiplexers are controlled by the microcontroller. An instrumentation amplifier follows the MUX and amplifies the differential signal between the two leads for the bipolar case or difference signal between each selected unipolar lead and the Wilson lead. The amplified signals are filtered by a second order high-pass (HP) Butterworth filter with cutoff frequency at 0.5Hz to eliminate the dc component. A fourth order low-pass Butterworth filter comes next with a cutoff frequency at 40Hz to eliminate the high-frequency noise. The line noise of 50Hz is rejected by a special band reject (BR) filter. The final step is to amplify the filtered signal and to add a dc component to create strictly positive signal values for the microcontroller [5].

The three analog signals (temperature, bipolar leads and unipolar leads) are converted to digital, by a 10 bit Analog to Digital Converter embedded in the development board Arduino Yún. The Arduino Yún contains a microcontroller ATmega 32u4 running at 16MHz and a microprocessor Atheros AR9331 running at 400MHz accompanied by a LINUX distribution. Furthermore, the Arduino Yún provides wired (Ethernet) and wireless (Wi-Fi) capabilities, as well as a Micro-SD card slot [6].

The microcontroller was programmed in C++. In order to calculate the heart rate from the bipolar leads, a sample of 3 seconds is taken to establish thresholds in order to identify the R wave which in turn signals a full heart cycle. The temperature is calculated using the equation from the thermistor calibration every 15 seconds. Furthermore, the microprocessor acts as a web server allowing the real time monitoring of the heart rate and temperature by means of web pages with embedded JavaScript code. The web pages also allow the selection of leads. Finally, the device supports the storing of acquired measurements in a micro-SD card, 2GB capacity, or in a Google spreadsheet.

3. Results and Discussion

Figure 2 presents the external and the internal view of the device.

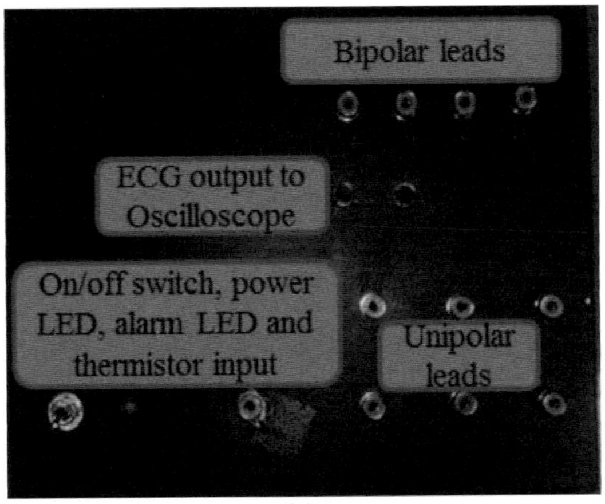

(a)

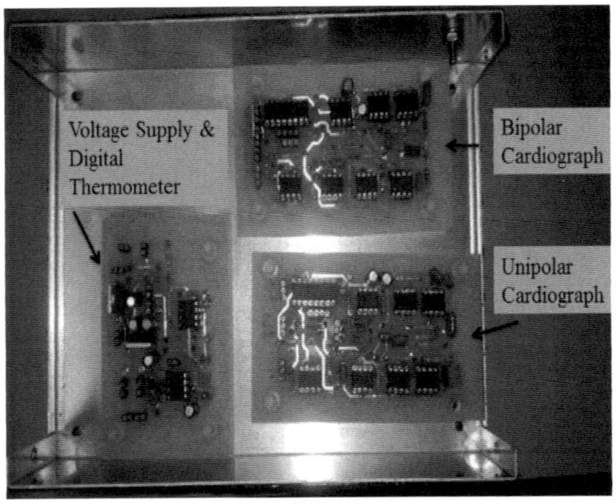

(b)

Figure 2. (a) The outside view of the device. (b) The inside view without the jumper connectors.

An ECG simulator was used to determine the accuracy of calculating the beats per minute (BMP) under various conditions (no noise, line noise, tremor, random noise and baseline wander) (Table 1).

Table 1. Measurements of beats per minute (BPM) for different types of noise.

Real BPM	No noise	Line 50Hz	Tremor	Random	BL.Wander[a]
30	30	30	30	30	30
50	50	50	50	50	50
75	75	75	75	75	75
100	100	102	99	100	99
130	128	131	130	130	131
150	149	150	150	151	149
200	202	199	202	197	202

[a] Base Line Wander. A low-frequency noise that can be in the frequency bandwidth of interest.

The calculation of the heart rate allows an error of ±1 bpm, according to many protocol tests [7] or a maximum of ±2 bpm according to many manufacturers [8]. In our case, a maximum error of ±3 bpm was observed.

The digital and the Hg thermometer have an accuracy of ±0.1°C. According to many manufacturers, an error of 0.1°C is often permitted, but also errors up to 0.3°C can be acceptable [9]. The percentage of those errors is often depending on the type of using medical thermistor (YSI 400, YSI 700) and the accuracy of reference thermometer. In our case, the difference between reference and calculated values has a maximum error of 0.3°C (Table 2).

Table 2. Measurements of testing temperatures in pairs.

Hg Thermometer (Reference Values)	Temperature Readings	Commercial Digital Thermometer (Reference Values)	Temperature Readings
26.7	26.6	36.8	36.6
32.7	32.9	36.6	36.9
37.0	37.2	36.8	36.6
41.4	41..3	36.4	36.7

4. Conclusion

In this paper a custom made portable device that measures a subset of vital signs was presented and can be used to home care and in hospitalization. The device provides fairly accurate results regarding the measurement of heart rate, under various noise conditions, and temperature readings for a cost of 160€ to 180€. The device could be made in smaller form factor using SMD technology and could be enriched by incorporating measurements of respiratory rate, blood pressure and oximetry. Also the Arduino Yún could be programmed to act as a web server providing worldwide access through the web pages.

References

[1] GE Healthcare, 2007, Datex - Ohmeda S/5 Compact Anesthesia Monitor, p 37, 48, 49, 53, 55, 59, 61, 63, 67, 69, 70, 71, 73, 77, 81 - 83.
[2] GE Healthcare, DINAMAP ProCare Monitor - Service Manual, 2008, p A-4, A-5, A-7, A-9.
[3] Roteriu C, Costin H, Arotaritei D and Dionisie B 2008 *A Wireless ECG Module for Patient Monitoring Network* Technological University "Gheorghe Asachi" **LIV** LVIII
[4] Delano M K 2012 A Long Term Wearable Electrocardiogram (ECG) Measurement System *Massachusetts Institute of Technology*
[5] Clark J W Jr, Neuman M R, Olson W H, Peura R A, PrimianoF P, Sideband M P, Webster J G and Wheeler L A 2001 *Medical Instrumentation, Application & Design 3rd Edition* ed J G Webster (John Wiley & Sons, Inc)
[6] http://www.arduino.cc/en/Main/ArduinoBoardYun first accessed in 14/04/2014
[7] Rigel Medical 2012 *An Introduction to Measuring and Simulating Vital Signs Version 1.2* p 23
[8] Nihon Kohden *Technical Data Vismo Bedside Monitor PVM-2701/2703* p 3
[9] GE Healthcare 2007 *Solar™ 8000M/I Patient Monitor* Operator's Manual Software Version 5.0 p H-5

Doses to medical workers operating in a PET/CT department after the use of new dynamic techniques.

K Dalianis[1],G Kollias[1], J Malamitsi[2], R Euthimiadou[3], J Andreou[3], E Georgiou[3], V Prassopoulos[3].

[1]Medical Physics Department, Hygeia SA, Athens, Greece

[2]Medical Physics Laboratory School of Medicine National and Kapodistrian University of Athens, Greece

[3]PET/CT Department, Hygeia SA, Athens, Greece

E-mail: kdalianis@hygeia.gr
Konstantinos Dalianis MSc, PhD, Hygeia Hospital, Erythrou Stavrou 4 & Kifissias P.O, 15123 Marousi, Athens, Greece

Abstract: Since new radiopharmaceuticals are used like [18F]-fluoro-3'-deoxy-3'-L-fluorothymidine and 18F fluoromethylcholine, also new dynamic techniques of imaging are used, measurements concerning the doses to medical staff are needed. The aim of this study was to measure the effective whole body dose of the personnel and compare them with the oldest. Estimation of equivalent dose for all members of the staff was monitored with the use of TLDs badges and electronic dosimeters. The duration of the study was year 2011 (983 patients).Concerning the nurses, we measured 10% increase in the wholebody doses and that is due to the longer time they spent near the patient (dynamic protocol). For technologist we measure 15-21% increase for they come near the patient immediately after administration. We can observe that there is an increase of the doses for technologists and nurses the numbers are significantly lower than the recommended annual dose limit by Euratrom 97/43.

Introduction:

The PET/CT applications have been continuously increasing for diagnostic procedures. Although such an increase is a positive trend for the benefit of patients, the associated risk of radiation exposure of staff needs to be properly evaluated.

Positron Emission Tomography is considered to be one of the most relevant diagnostic imaging techniques having peculiar characteristic to provide both functional and morphological information for the patient. The most commonly used radiopharmaceutical is 18 F- Fluorodeoxyglucose (18F-FDG). [1]

Due to the high-energy tracers emitting 511 KeV and considering the risks associating to ionizing radiation that have been derived from the past, special attention is needed when dealing with radiation protection aspects in a PET/CT facility [2]. Medical workers are exposed to ionizing radiation during the procedure of PET/CT. Since new radiopharmaceuticals are used like [18F]-fluorothymidine (FLT) and 18F fluoromethylcholine (FCH) new techniques of imaging are used and new measurements concerning the doses to medical staff are needed [3]. Different activities of F-18 FDG, FLT and FCH are used in our center [4].

The aim of this study was to measure the effective whole body dose of the personnel and compare them with the oldest measurements, a comparison before and after the use of new dynamic techniques. Our department was the first PET/CT department operated in Greece in 2004, and the authors investigated and recorded the daily radiation exposure of the staff since the first day of operation. [1]

Material and Method:

The estimation of equivalent dose from external dosimetry for all seven (7) members of the staff (2 nurses, 2 medical physicists, 2 technologists, and 1 doctor) was monitored with the use of TLDs

(TLD-100, LiF:Mg,Ti) badges and electronic dosimeters worn at the upper pocket of their overall photo 1. The average workload of the department is 6-8 patients per day [5].

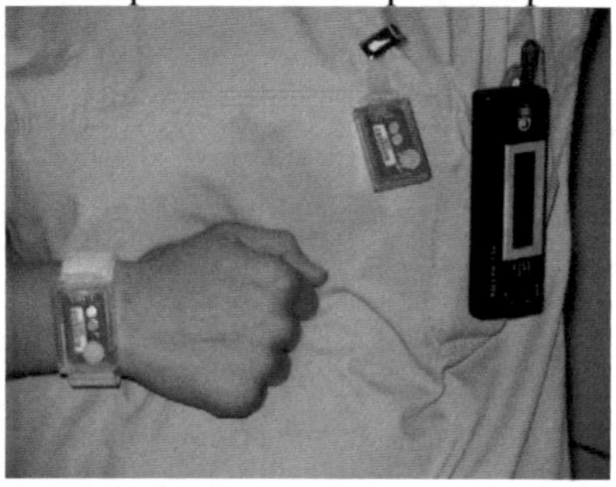

Photo 1: TLDs badges and electronic dosimeters.

During the time of the study 983 patients were examined. In our department 18F-FDG, 18F-FLT, 18F-FCH is available in multi dose vials. The vial arrives to the department in a 7 cm thick lead shielded container and it is placed in the hot lab. The lead container is brought in the lead hood, a tailor made hood with 20cm lead shield. With the use of a tong the vial is been placed in the dispenser (15cm of lead) [6].

The radiopharmaceuticals used for imaging were obtained with the use of shielded dispenser system which includes a mirror at the top in order for the multidose vial to be fully covered.

At all time, one nurse, one medical physicist, one technologist and one doctor are operating in the department. [7]

Nuclear Medicine procedures require patient interaction relating to patients preparation administration of radioactive and non-radioactive medication or parental route, explaining the procedure comforting and reassuring the patients. [3].

The typical 18F-FDG procedure for a patient that undergoes a whole body scan is the following:

Patient is injected with the radiopharmaceutical lying in a bed in the waiting area and he/she is asked to wait for 50 minutes. Before the beginning of the examination, the patient is asked to empty the bladder. The PET/CT examination last for about 35 minutes and then the patient, after changing, leaves the department. [8]

On the contrast, when 18F-FLT or 18F-FCH procedures are performed, the injection takes place inside the PET/CT room, were a dynamic examination is been performed. Specifically 3 min acquisition data are collected for 30 min and afterwards one bed of 12 min duration is held.

Regarding the responsibilities for each worker, Medical Physicist is handling the multidose vial and the dose segmentation. The nurses involved in this study are responsible for the dose administration. Also, nurses monitor the patient through the 50 minute uptake procedure after the injection in which time they provide assistant to the patient if needed. When the FLT or FCH procedure is performed the injection is done inside the PET/CT room. [9]

The technologist duties include escorting the patient in the examination room and positioning the patient in the examination table. In cases of FLT or FCH procedures were dynamic examination is been performed, the technologist interacts with higher amount of radioactivity, since they are near to the patient at the exact time of the injection and not after the 50 min uptake time like in FDG examinations [10].The technologist is also responsible for helping the patient after the end of the examination.

Data were collected day-to-day concerning the interaction of the staff during PET/CT procedures. The technologists and nurses that are operating in our PET/CT facility have experience in Nuclear Medicine procedures for more than 15 years and are trained and experienced interacting with radiopharmaceuticals [11]. During the course of this study the actions of each staff member were observed and recorded. Medical Physicists were responsible for collecting the data concerning the time spent by the staff at less than 2 meters from the radioactive source or from the syringe containing the radiopharmaceutical or from the injected patient, as well as patient data concerning the administered activity and the patient status [1].

Results:

We compared the first six months of 2011 a period in which no FLT or FCH procedures were performed, with the second semester in which 50 FLT and 26 FCH examinations were performed. The average of FDG patients was the same for every day, and all patients, FDG-FLT- FCH received 330-390 MBq dose each [4].

The time spent near open sources was 1-3 min for the nurses and 6-9 min for technologist depending on patient mobility and character. [12]

Dose values of the PET/CT staff were collected monthly with the use of TLDs dosimeters that are measured from the Greek Atomic Energy Commission and daily with the use of electronic dosimeters that were recorded by the radiation safety officer (Medical Physicists). [3]

The whole body dose for the first six months for the medical workers were the following:

Nurse #(1) received 1,3 mSv as a whole body dose and Nurse #(2) received 1,1 mSv respectively. In figure 1 we can see the monthly whole body dose for Nurse (1) and (2).

For Medical Physicist #(1) the dose value was 1,9 mSv and for Medical Physicist #(2) was 2,2 mSv respectively. Figure 2 saws the monthly dose distribution to medical physicist.

Lastly, Technologist #(1) received 1,1 mSv as a whole body dose and Technologist #(2) received 1,4 mSv respectively.

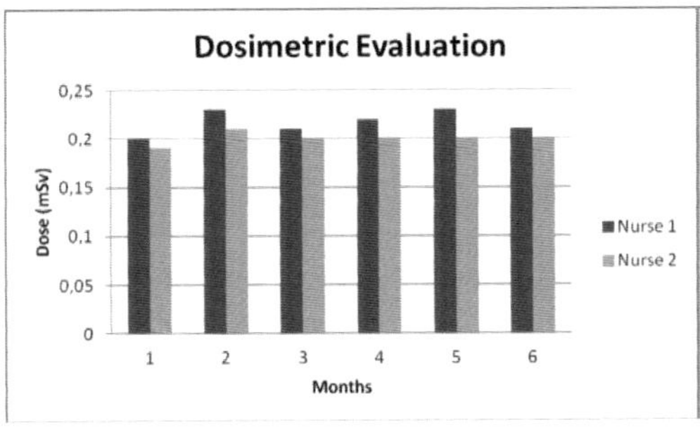

Figure 1, Dosimetric evaluation of the nurses for the first semester.

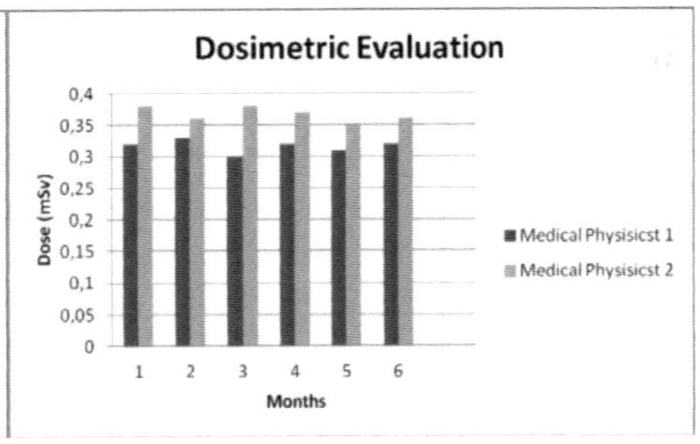

Figure 2, Dosimetric evaluation of the medical physicist for the first semester

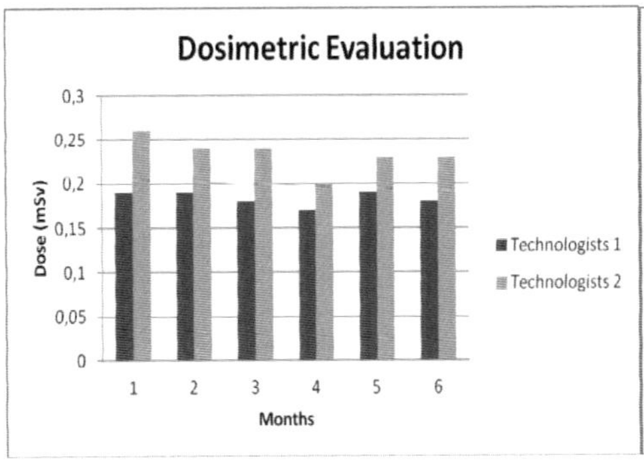

Figure 3, dosimetric evaluation of the technologist for the first semester.

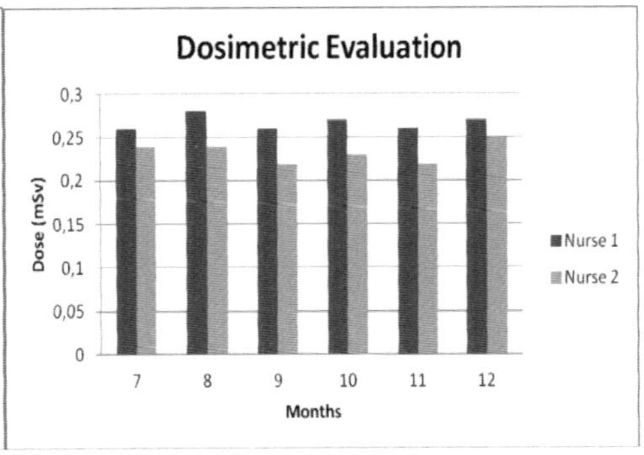

Figure 4, Dosimetric evaluation of the nurses for the second semester.

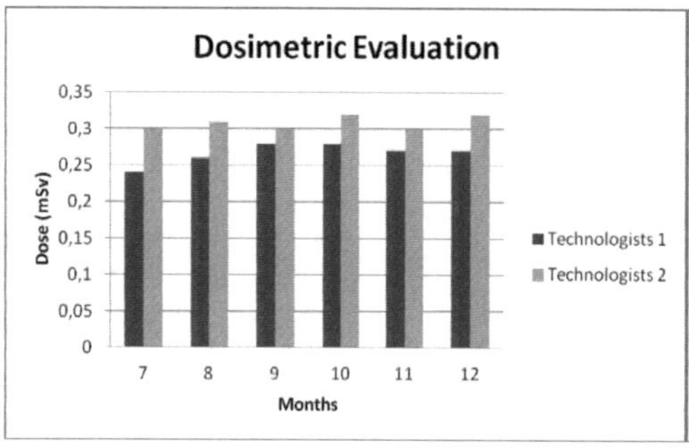

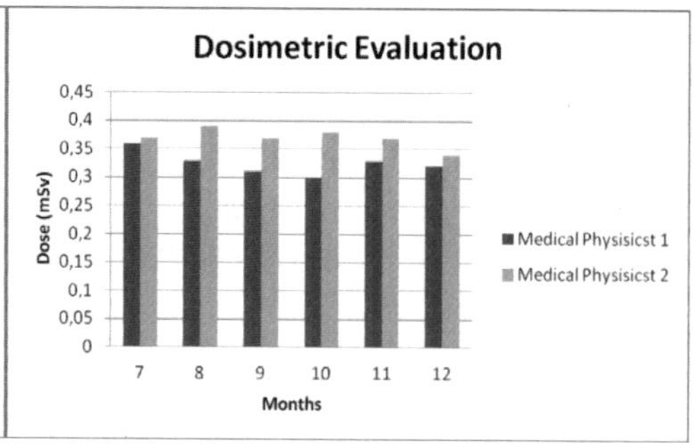

Figure 5, Dosimetric evaluation of the technologist for the second semester. Figure 6, Dosimetric evaluation of the medical physicist for the second semester

After the use of FLT and FCH at the second semester, the doses were:
Nurse #(1) received 1,6 mSv and Nurse #(2) received 1,4 mSv.
Medical Physicist #(1) received 1,95 mSv and Medical Physicist #(2) 2,19 mSv.
Technologist #(1) received 1,6 mSv, Technologist#(2) received 1,8 mSv.
In figures 4, 5, 6 we can see the monthly distribution after the use of FLT and FCH.
Regarding the whole body doses for medical physicist we can see that there were no changes and that is because they do not interact with patients undergoing FLT and FCH procedures. Concerning the nurses presenting the quantitive correlation we measured a 17-20% (per cent) increase in the whole body doses and that is due to the longer time they spent near the patient, since FLT and FCH examinations interact to the patient for longer time. For the technologist we measure a 15-21% (per cent) increase and that is because they come in touch with the patient, not after 50 min like in FDG procedures, but they are near to the patient at the time of the injection.

Conclusions

From our results we can observe that although there is an increase of the doses for technologists and nurses the numbers are significantly lower than the recommended annual dose limit by Euratrom 97/43.

References

[1] Dalianis K, Malamitsi J, Gogou L, Pagou M, Efthimiadou R, Andreou J, Louizï A, Georgiou E Dosimetric evaluation of the staff working in a PET/CT department, *Nuclear Instruments and Methods in Physics Research A* **569**, 548 – 550.
[2] Biran T, Weininger J, Malchi S, Marciano R, Chisin R. Measurements of occupational exposure for technologist performing 18F FDG PET scans. *Health Phys* 2004; **87**:539-44
[3] Amaral A, Itie C, Bok B, Dose Absorbed by Technicians in Positron Emission Tomography Procedures with FDG, *Brazilian Archives of Biology and Technology* **50**: pp.129-134, Semptember 2007.
[4] Everaert H, Vanhove C, Lahoutte T, Muylle K, Caveliers V, Bossuyt A, Franken PR Optimal dose of 18-FDG required for whole body PET using LSO PET camera *Eur J Nucl Med Mol Imagining* 2003;**30**:1615-9. Epub 2003 Sep 23.
[5] Roberts FO, Gunawardana DH, Pathmaraj K, Wallance A, U PL, Berlangieri SU, Radiation dose to PET technologists and strategies to lower occupational exposure. *J Nucl Med Technol.* 2005; **33**: 44-7.
[6] Khan SH, Ahmad SM, Whole Body Dose to Radiation Workers at a Tertiary Care Hospital in Kashmir: Analysis and Comparison with Naional Multi-Institutional Pool of Medical Radiation Workers, *IJNM*, **18**(3): 57-61, 2003.
[7] Seierstad T, Stranden E, Bjering K, Evensen M, Holt A, Michalsen HM, Wetteland O, Doses to Nuclear Technicians in a dedicated PET/CT centre utilising 18F Fluorodeoxyglucose (FDG), *Radiation Protection Dosimetry* February 2007 **123** (2): 246-249.
[8] Benatar NA, Cronin BF, O´Doherty MJ Radiation dose rates from patients undergoing PET: implications for the technologist and waiting areas *Eur J Nucl Med* 2000;**27**:583-9.
[9] V Vijayakumar, E Briscoe, S Vijayakumar. Role of Nurses in Daily Nuclear Medicine. *The Internet Journal of Nuclear Medicine.* 2006 **Volume 3** Number 2.
[10] Zeff PB, Yester MV, Patient self-attenuation and technologist dose in positron emission tomography. *Med Phys* 2005;**32**:861-5.
[11] Lundberg TM, Gray PJ, Bartlett ML, Measuring and minimizing the radiation dose to nuclear medicine technologists. *J Nucl Med Techol.* 2002;**30**:25-30.
[12] Buchillier T, Aroura A, Bochud F, Schuler C, Stritt N, Valley JF, Wernli C. Calibration and testing of a TLD dosimeter for area monitoring, *Radiat. Prot Dosimetry* 2004; **110**:1173-9.

X-ray Luminescence Efficiency of GAGG:Ce Single Crystal Scintillators for use in Tomographic Medical Imaging Systems

S L David, I G Valais, C M Michail and I S Kandarakis

Radiation Physics, Materials Technology and Biomedical Imaging Laboratory, Department of Biomedical Engineering, Technological Educational Institute of Athens, Egaleo, 122 10 Athens, Greece

email: sdavid@teiath.gr

Abstract. The purpose of the present study was to evaluate different scintillator crystal samples, with a cross section of 3×3mm^2 and various thicknesses ranging from 4mm up to 20mm, of the new mixed $Gd_3Al_2Ga_3O_{12}$:Ce (GAGG:Ce) scintillator material under X-ray irradiation, for potential applications in Tomographic Medical Imaging systems. Evaluation was performed by determining the X-ray luminescence efficiency (XLE) (emitted light energy flux over incident X-ray energy flux) in energies employed in general X-ray imaging. For the luminescence efficiency measurements, the scintillator samples were exposed to X-rays using a BMI General Medical Merate tube, with rotating Tungsten anode and inherent filtration equivalent to 2 mm Al. X-ray tube voltages between 50 to 130 kV were selected. An additional 20 mm filtration was introduced to the beam to simulate beam quality alternation equivalent to a human body. The emitted light energy flux measurements were performed using an experimental set up comprising a light integration sphere coupled to an EMI 9798B photomultiplier tube which was connected to a Cary 401 vibrating reed electrometer. The GAGG:Ce sample with dimensions 3×3×10 mm^3 exhibited higher XLE values, in the whole X-ray energy range examined. XLE value equal to 0.013 was recorded for this crystal at 130 kVp - a setting frequently used in Computed Tomography applications.

1. Introduction

Scintillators are used as radiation converting media in various applications, from medical imaging to high energy physics experiments [1-4]. Particularly, in medicine, scintillators have been applied from low energy examinations, such as mammography and general radiography to nuclear medicine and radiotherapy, in powder or crystal form [5]. Among the various crystal detectors, Cerium (Ce) doped scintillators are the dominant due to their internal properties, such as, fast emission and high light yield [2,5]. Ce doped Gadolinium Aluminum Gallium Garnet ($Gd_3Al_2Ga_3O_{12}$:Ce or GAGG:Ce) is a recently developed mixed scintillator crystal with high density (6.69 gcm^{-3}), fast scintillation response (~90 ns) and high light yield (~46000 ph/MeV) [6]. Compared to many current efficient scintillators, GAGG:Ce is non hygroscopic (like $LaBr_3$:Ce, $LaCl_3$:Ce, CsI and NaI:Tl crystals) and it does not have natural radioactivity (like Lutetium based materials) [6-8]. GAGG:Ce emits optical photons at 520nm and has an effective atomic number equal to 54.4. As it has been previously reported [7], the production of GAGG:Ce is based on the Czochralski method. Several studies of energy resolution and timing performance of GAGG crystals readout, using a variety of photon detection technologies

(APD, SIPM, PMT, MPPC, PIN diode), have shown that the crystal is promising for medical imaging and gamma-ray spectroscopy applications [9-10]. Moreover, the development of a sub millimeter GAGG block detector, using 0.4 mm GAGG pixels, was reported for PET and SPECT applications [11]. Very recently, the evaluation of the GAGG:Ce scintillator array (with 2×2×5 mm^3 pixel elements) under low energy isotopes (^{57}Co and ^{99m}Tc) for SPECT applications, as well as, with ^{137}Cs and ^{22}Na isotopes for PET imagers based on 4×4 discrete SIPM optical detector were published [12-13].

Although several studies reported on the performance of the GAGG:Ce scintillator material under gamma ray excitation, only a few have been published for X-ray excitation [14], e.g. for potential use in tomographic X-ray imaging or integrated PET/CT and SPECT/CT detector systems.

The purpose of the present study was to evaluate seven different scintillator crystal samples with a cross section of 3×3mm^2 and various thicknesses, ranging from 4mm up to 20mm of the new mixed GAGG:Ce scintillator material under X-ray excitation for applications in tomographic medical imaging detectors, i.e. CT or integrated SPECT/CT or PET/CT detectors. Experimental measurements such as the X-ray luminescence efficiency (light energy flux over exposure rate) under X-ray excitation in the radiographic energy range from 50 to 130 kVp were performed.

2. Materials and Methods

GAGG:Ce single crystals are produced in the Materials Research Laboratory, Furukawa Co. Ltd. in Japan. The seven scintillators were cut from the same bolus with 4, 5, 6, 8, 10, 15 and 20 mm thicknesses and a cross section of 3x3mm^2. All the crystal samples surfaces have been polished. The dimension of the GAGG:Ce crystal's output face was selected to be 3x3mm^2, in order to have compatibility between scintillator and photodetector's pixel size, since most new optical detectors, that are currently used in medical imaging systems have similar pixel sizes [15-16].

The crystals were exposed to X-rays from a BMI General Medical Merate tube with rotating Tungsten anode and inherent filtration equivalent to 2 mm Al, with energies ranging from 50 to 130 kVp. An additional 20 mm filtration was introduced in the beam to simulate beam quality alternation similar to a human body [17]. The light flux measurements were performed using a light integration sphere (Oriel 70451), coupled to a photomultiplier (EMI 9798B), connected to a Cary 401 vibrating reed electrometer [18]. The X-ray exposure rate was measured at the crystal's position using a Radcal 2026C dosimeter.

2.1. X-ray luminescence efficiency (XLE)

The X-ray luminescence efficiency (XLE) describes the efficiency of a scintillator to transfer the signal from the input to the output. XLE can be determined by the ratio of the emitted light energy flux (Ψ_Λ) over the incident X-ray energy flux (Ψ_0): ($XLE = \Psi_\Lambda / \Psi_0$). XLE was determined by measuring the X-ray exposure rate and emitted light energy flux [18]. The X-ray energy flux (Ψ_0) was determined by converting X-ray exposure rate (X) [18], as follows: $\Psi_0 = X\hat{\Psi}$ where $\hat{\Psi}$ is defined as the X-ray energy flux per exposure rate, estimated according to (1):

$$\hat{\Psi} = \int \Psi_0(E)dE / (\int \Psi_0(E)\left[X / \Psi_0(E)\right]dE) \tag{1}$$

where

$$X / \Psi_0(E) = (\mu_{en}(E) / \rho)_{air} \cdot (W_A / e)^{-1} \tag{2}$$

is the factor converting energy flux into exposure rate, $(\mu_{en} / \rho)_{air}$ is the X-ray mass energy absorption coefficient of air, at energy E, and W_A / e is the average energy per unit of charge required to produce an electron-ion pair in air. W_A / e and $(\mu_{en} / \rho)_{air}$ were obtained from tabulated data [19].

2.2. Detector quantum optical gain (DQG)

Detector quantum optical gain (DQG) is the ratio of the light photon flux Φ_Λ over the X-ray photon flux (Φ_X). Using this quantity, the emitted light photon flux can be expressed in terms of experimentally measurements (absolute efficiency, exposure rate, mean light wavelength), by using (3):

$$\Phi_\Lambda = \Psi_\Lambda / hc\bar{\lambda}^{-1} \qquad (3)$$

Where the denominator is the mean energy of the emitted light photons, λ being the mean light wavelength determined from emission spectra measurements. Φ_X was determined by using (1) and (2), replacing Ψ_0 by Φ_0 and dividing (2) by the X-ray energy [18]. More details about the quantity of DQG can be found in [20].

3. Results and Discussion

Figure 1 shows the XLE for the seven GAGG:Ce crystals, for various tube voltages. In every case, XLE values show a tendency to decrease with increasing tube voltage, due to the energy absorption properties and light transmission through the crystals. The higher XLE values are shown for the 3×3×10 mm^3 GAGG:Ce crystal. Crystals with higher thickness presents slightly lower XLE values due to increase of the scintillation light absorption by the crystal mass.

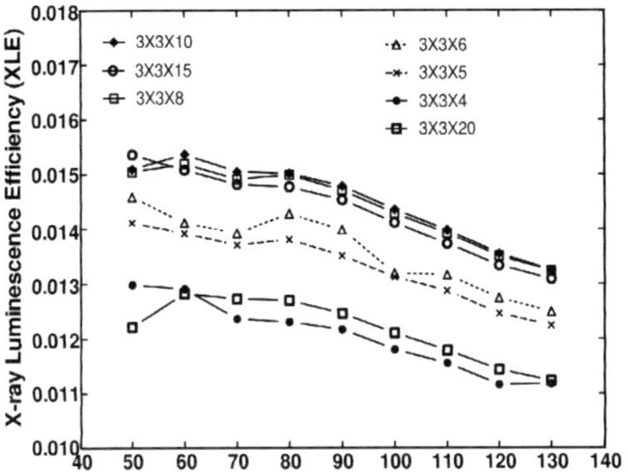

Figure 1. Variation of the X-ray luminescence efficiency (XLE) of the GAGG:Ce crystals with X-ray tube voltage.

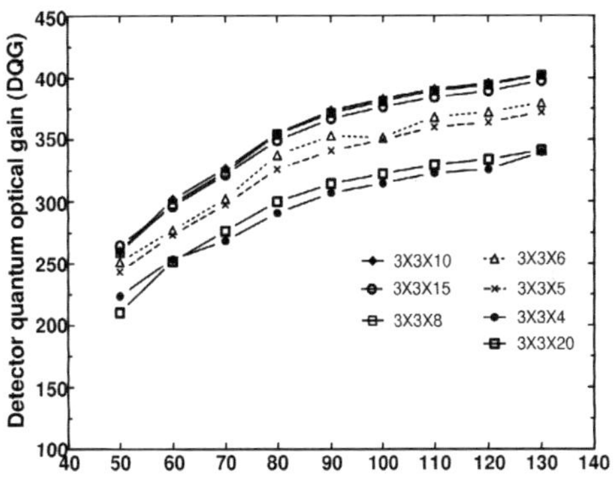

Figure 2. Variation of the detector quantum gain (DQG) of the GAGG:Ce crystals with X-ray tube voltage.

Figure 2 show DQG values for the GAGG:Ce crystals, in the energy range under investigation. DQG can be a measure of light yield and increases as the photon energy increase, showing that GAGG:Ce crystals could produce efficiently optical photons up to the X-ray computed tomography (CT) energy range (i.e. 130kVp). DQG reflects more accurately the detector intrinsic properties, since it is not affected by the air-ionization chamber properties used to determine AE. The latter depends on the X-ray energy flux to exposure conversion factor, which exhibits a highly non-linear variation with X-ray energy. The behaviour of DQG curves can be explained by considering the following: The number of optical photons created per absorbed X-ray increases due to the light created per photon is larger in higher X-ray energies and therefore DQG increase. We note that the X-ray photons are totally absorbed by the GAGG:Ce scintillators mass, due to high density of the material.

4. Conclusions

In the present study, seven different GAGG:Ce scintillator crystals with a cross section of $3\times3mm^2$ and various thicknesses were examined under X-ray irradiation for potential applications in tomographic medical imaging detectors. The GAGG:Ce sample with dimensions equal to $3\times3\times10$ mm^3 provided higher XLE and DQG values, almost in the whole X-ray energy range examined. XLE value equal to 0.013 and DQG up to 400 was recorded for this crystal at an X-ray tube voltage of 130 kVp used in Computed Tomography applications.

Acknowledgments
This research has been co-funded by the European Union (European Social Fund) and Greek national resources under the framework of the "Archimedes III: Funding of Research Groups in TEI of Athens" project of the "Education & Lifelong Learning" Operational Programme.

References
[1] Blasse G G and Grabmaier B C, 1994, *Luminescent Materials* (Springer-Verlag, Berlin, Heidelberg)
[2] Van Eijk C W E 2002 *Phys. Med. Biol.* **47** 85
[3] Kandarakis I S 2014 *J. Lumin.* in press: http://dx.doi.org/10.1016/j.jlumin.2014.11.009
[4] Guerra A 2004 *Ionizing Radiation Detectors for Medical Imaging*, (World Scientific Publishing Company, Singapore) p 53
[5] Nikl M 2006 *Meas. Sci. Technol.* **17** 37
[6] Iwanowska J *et al.*, 2013 *Nucl. Instr. and Meth. Phys. Res. A.* **712** 34
[7] Kamada K, Yanagida T, Endo T, Tsutumi K, Usuki Y, Nikl M, Fujimoto Y, Fukabori A and Yoshikawa A 2012 *J. Cryst. Growth* **352** 88
[8] Kobayashi M, Tamagawa Y, Tomita S, Yamamoto A, Ogawa I and Usuki Y, 2012, *Nucl. Instr. and Meth. Phys. Res. A.* **694** 91
[9] Yeom J Y, Yamamoto S, Derenzo S E, Spanoudaki V C, Kamada K, Endo T and Levin C S 2013 *IEEE Trans. Nucl. Sci.* **60** 988
[10] Kato T *et al.*, 2013 *Nucl. Instr. and Meth. Phys. Res. A.* **699** 235
[11] Kataoka J *et al.*, 2013, *Nucl. Instr. and Meth. Phys. Res. A.* **732** 403
[12] Georgiou M, David S, Fysikopoulos E and Loudos G 2013 *IEEE Nucl. Sci. Symp.& Med. Imag conf.* M21-48 DOI: 10.1109/NSSMIC.2013.6829327
[13] David S, Georgiou M, Fysikopoulos E and Loudos G 2015 *EJMP* DOI: http://dx.doi.org/10.1016/j.ejmp.2015.03.008 in press
[14] Seferis I E, Michail C M, David S L, Bakas A, Kalivas N I, Fountos G P, Panayiotakis G S, Kourkoutas K, Kandarakis I S and Valais I G 2014 *IFMBE* proc. **41** 455
[15] http://sensl.com/products/silicon-photomultipliers/
[16] http://www.hamamatsu.com/us/en/product/category/3100/4004/4113/index.html#
[17] Michail C, Kalyvas N, Valais I, David S, Seferis I, Toutountzis A, Karabotsos A, Liaparinos P, Fountos G and Kandarakis I 2013 *J. Lumin.* **144** 45
[18] Michail C, Valais I, Seferis I, Kalyvas N, David S, Fountos G and Kandarakis I 2014 *Radiat. Meas.* **70** 59
[19] Hubbell J H, Seltzer S M, 1995, Tables of X-ray mass attenuation coefficients and mass energy absorption coefficients 1 to 20MeV for elements Z=1 to 92 and 48 additional substances of dosimetric interest. US Department of commerce, NISTIR 5632
[20] Kalyvas N, Valais I, David S, Michail C, Fountos G, Liaparinos P and Kandarakis I 2014 *Optics and Spectroscopy* **116** 743

Design considerations for a C-shaped PET system, dedicated to small animal brain imaging, using GATE Monte Carlo simulations

N Efthimiou[1], P Papadimitroulas[2], T Kostou[2], G Loudos[1]
[1] Department of Biomedical Engineering, Technological Educational Institute of Athens, Greece.
[2] Department of Medical Physics, School of Medicine, University of Patras, Greece.

E-mail: efthimioun@upatras.gr

Abstract. Commercial clinical and preclinical PET scanners rely on the full cylindrical geometry for whole body scans as well as for dedicated organs. In this study we propose the construction of a low cost dual-head C-shaped PET system dedicated for small animal brain imaging. Monte Carlo simulation studies were performed using GATE toolkit to evaluate the optimum design in terms of sensitivity, distortions in the FOV and spatial resolution. The PET model is based on SiPMs and BGO pixelated arrays. Four different configurations with C-angle 0°, 15°, 30° and 45° within the modules, were considered. Geometrical phantoms were used for the evaluation process. STIR software, extended by an efficient multi-threaded ray tracing technique, was used for the image reconstruction. The algorithm automatically adjusts the size of the FOV according to the shape of the detector's geometry. The results showed improvement in sensitivity of ~15% in case of 45° C-angle compared to the 0° case. The spatial resolution was found 2 mm for 45° C-angle.

1. Introduction

Nowadays, most clinical and preclinical PET scanners rely on the cylindrical or near-cylindrical geometry [1-5]. Yet, there are some implementations which are based on alternative geometries such as rotating [6-9] or static planar heads [10-12]. The incorporation of SiPM arrays provides the means to design flexible PET scanners so as to improve sensitivity, with limited number of detector heads [13]. The design and configuration, of such systems, strongly depends on the particular region under investigation, but the driving factors are to maximize sensitivity and lower the cost [14].

In this study we evaluated the performance of a design for a C-shaped PET scanner designed for small animal brain imaging. We tried to incorporate the minimum amount of detectors, in order to keep the cost as low as possible. The designs were evaluated in terms of sensitivity, distortions in the FOV and spatial resolution.

2. Material and Methods

2.1. GATE simulations

We used the GATE (v.7) simulation toolkit to perform the simulations [15]. The geometry was based on a realistic hypothesis of six modules of a 6×6 BGO array [16], with crystal size $2.0\times2.0\times0.5$ mm^3

coupled to SiPM arrays (1.35×1.35 mm^2). Each, head is assembled from 3 modules, which can be tilted, with respect to the central module, forming the C-shape. The head separation distance was defined as the distance between the central detector modules of each head. In this experimental setup was set to 3 cm.

In order to evaluate the effect of the C-angle to the performance of the system, four configurations of 0°, 15°, 30° and 45° C-angle, were considered. The configurations are illustrated in Figure 1. The size of the FOV and thus of the reconstructed images depends on the C-angle and was 59×59, 52×52, 48×48 and 44×44 pixels for 0°, 15°, 30° and 45° C-angle, respectively.

After the simulations a 400-650keV energy window, was applied. Random and scattered corrections were not considered.

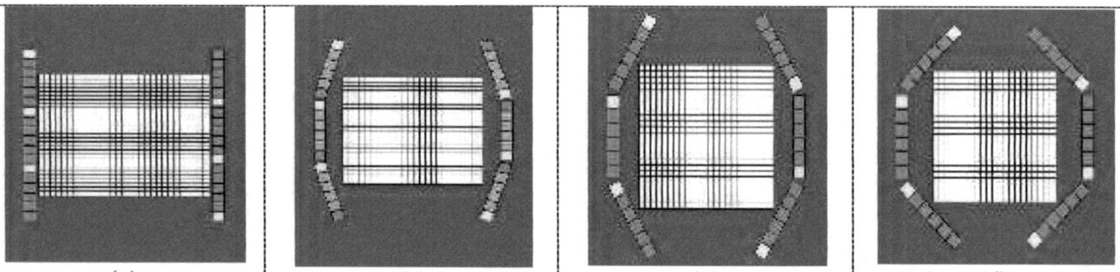

(a) (b) (c) (d)

Figure 1. Illustration of the system configurations. (a) 0° (planar imaging), (b) 15°, (c) 30° and (d) 45° C-angle.

2.2. Phantoms
In order to evaluate the system's performance different phantoms were simulated. For the evaluation of the sensitivity and the FOV distortions two cylindrical water phantoms, were used. The phantoms were located in the center of the FOV and had diameters of 5 and 10 mm with a 1 MBq ^{18}F ion source.

For the evaluation of the spatial resolution we simulated capillary sources, with diameter of 1.1 mm, in an "L" arrangement. The location of the capillaries was at the center of the FOV (cap 1), 5 mm displaced on the X axis (cap 2) and 5 mm on the Y axis (cap 3). Each capillary was coupled to an ^{18}F source of 150 KBq. Other authors have presented different evaluation methods [17].

2.3. STIR reconstruction
The reconstructions were performed with STIR image reconstruction toolkit [18]. STIR does not support non-cylindrical geometries, out of the box. We had to calculate the system probability matrix (SPM) in advance and provide it as input to STIR during reconstruction. A modified Siddon ray tracer was implemented for this purpose [19].

For this initial investigation, the detectors were considered black, although the starting and ending point of the rays could be located in random places inside the crystal's volume. No inter-crystal scattering was taken into account.

3. Results and Discussion

3.1. Count rate performance
For the evaluation of the count rate performance of the system the cylindrical phantom, with diameter 10 mm, was used. The results showed (Figure 2) that the performance is improved with the C-angle, because the angular coverage is increased. From 0° to 45° the increase is approximately 15%.

3.2. FOV Distortion
In order to investigate the effect of the limited angular sampling the cylindrical phantoms with diameter 5 and 10 mm and C-angles 0° and 45°, were used. The results showed that in the

reconstructed images both phantoms were reconstructed distorted in some extent [20]. Along the X axis were enlarged and on the Y axis are slightly suppressed. The magnitude of the distortion effect depends on the size of the phantom and the C-angle.

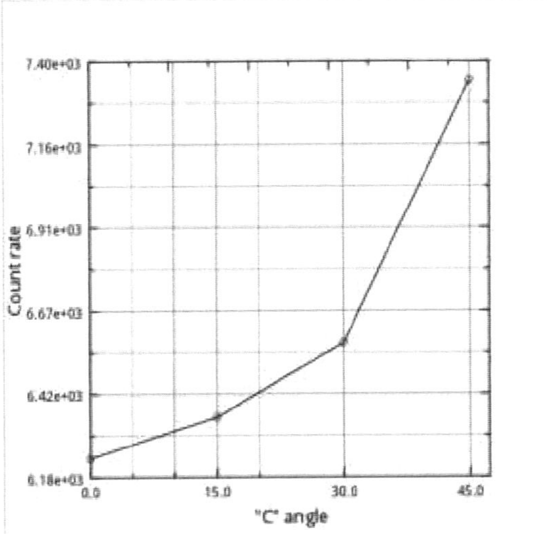

Figure 2. Count rate performance of the system over the C-angle.

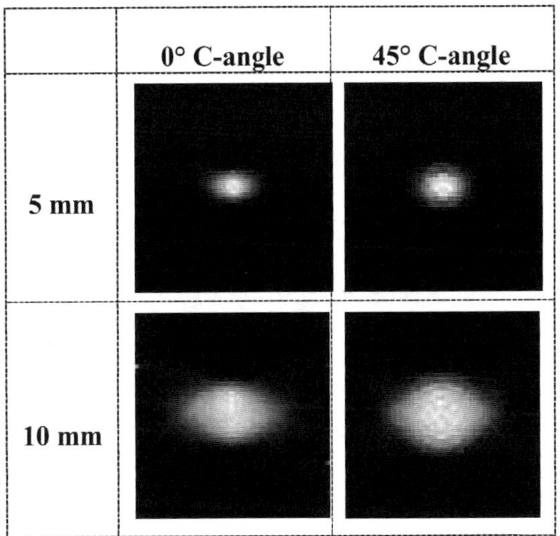

Figure 3. Reconstructed images of the two cylindrical phantoms with diameters 5 and 10 mm, for C-angles 0° and 45°

3.3. Spatial resolution

The spatial resolution results for 0° and 45 ° C-angle, are presented. The results showed that for the 0° C-angle the distortion of cap 2 interfered with cap 0, rendering the measurement on X axis impossible. On the Y axis the two sources are distinguished but are slightly deformed with profiles of 1.7 mm and 1 mm, for cap 1 and cap 3, respectively.

From the 45° configuration, due to the improved angular coverage, the spatial resolution in all three locations, is 2 mm, both on X and Y axis. The spatial resolution on planar systems strongly depends on the crystal size of the detectors.

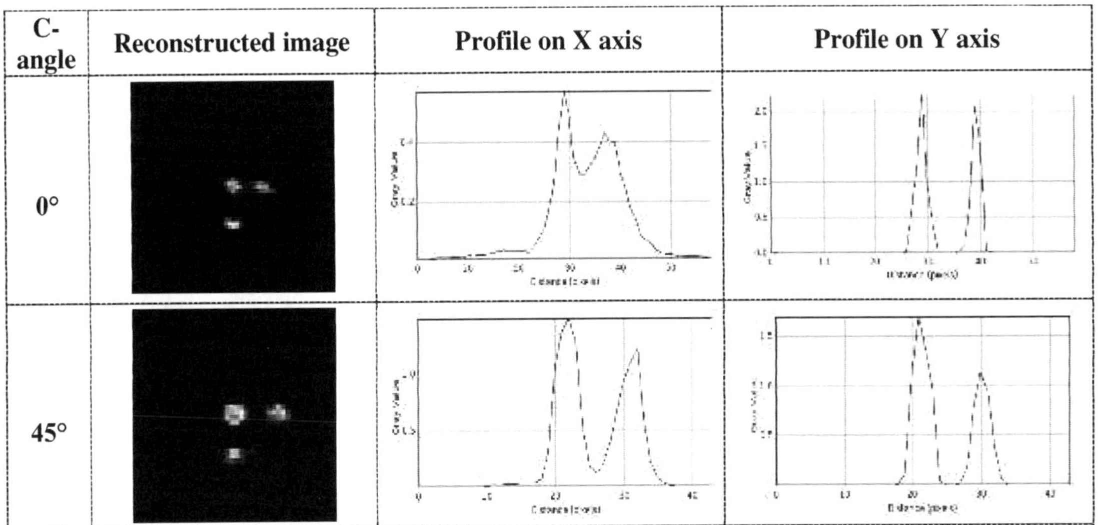

Figure 4. Reconstructed images after 4 iterations of three ^{18}F capillary sources, for C-angles 0° and 45° and the corresponding profiles on X and Y axis

4. Conclusions

Design considerations for a C-shaped PET system, dedicated to mouse brain imaging, were presented. The geometry was based on two opposite C-shaped heads. The C-angle of the heads was set to deferent values in order to investigate the impact on the sensitivity, distortions in the reconstructed image due to the limited angle sampling and spatial resolution.

It was found that the count rate performance of the system is improved for larger C-angles compared to planar head configuration, due to better angular coverage, on the expense of smaller images.

In addition, with larger C-angles the distortion in the FOV, due to the limited angle samples, is gradually reduced. The axis perpendicular to the heads is more sensitive to this effect. Other researched have also reported on this distortion and demonstrated that time of flight can eliminate them [20].

The spatial resolution of the system was found to be 2 mm, as the crystal size, when adequate angular sampling is achieved.

Acknowledgements
This research has been co-funded by the European Union (European Social Fund) and Greek national resources under the framework of the "ARISTEIA II" project HYPERGNOSTIC code 4309 of the "Education & Lifelong Learning" Operational Programme.

References

[1] Yu Ram A, Kim J S, Kang J H and Lim S M 2015 *J INST* **10**, P04001
[2] Kwon S I, Lee J S, Yoon H S, Mikiko I, Ko G B, Choi J Y et. al 2011 *J Nucl. Med.* **52**, 572
[3] Sanchez F, Orero A, Soriano A, Correcher C et. al. 2013 *Med. Phys.* **40** 051906
[4] Karpetas G E, Michail C M, Fountos G P, Kandarakis I S et. al. 2014 *Nucl. Med. Commun.* **35** 967
[5] Miranda-Menchaca A, Martinez-Davalos A, Murrieta-Rodriguez T, Sánchez-Alva H and Villafuerte-Rodriguez M 2015 *J INST* **10** T05008
[6] Islamirad S Z, Peyvandi R G, Lehdarboni M A and Ghafari A A 2015 *Nucl. Instr. Meth. Phys. Res. Sect. A* **781** 6
[7] Herraiz J L, Espana S, Cal-Gonzalez J, Vanquero J J, Desco M and Udias J M 2010 *Nucl. Instr. Meth. Phys. Res Sect A* **648** S169
[8] Efthimiou N, Maistros S, Tripolitis X, Samartzis A, Loudos G and Panayiotakis G 2011 *Meas. Sci. Tech.* **22** 114010
[9] Shao Y, Sun X, Lou K, Zhu X R et. al. 2014 *Phys. Med. Biol.* **59** 3373
[10] Zhang H, Bao Q, Vu N T, Silverman R W et.al. 2011 *Mol. Imag. Biol.* **13** 949
[11] Peng H 2015 *Nucl. Instr. Meth. Phys. Res. Sect.* A **779** 39
[12] Kao C-M, Dong Y and Chen C-T 2008 *IEEE Nucl. Sci. Symp. Conf. Rec.* M06-123
[13] David S, Georgiou M, Fysikopoulos E, Efthimiou N et. al. 2011 *IEEE Nucl. Sci. Symp. Conf. Rec.* 1886
[14] Raylman R R, Majewski S, Weisenberger A G and Popov V 2001 *J. Nucl. Med.* **42** 960
[15] Sarrut D, Bardiès M, Boussion N and Freud N 2014 *Med. Phys.* **41** 064301
[16] Valais I, Michail C, David S, Liaparinos P, Fountos G, Paschalis T, Kandarakis I and Panayiotakis G 2010 *IEEE Trans. Nucl. Sci.* **57**(1):3
[17] Karpetas G, Michail C, Fountos G, Valsamaki P, Kandarakis I and Panayiotakis G 2013 *Hell. J. Nucl. Med.* **16**(2) 111
[18] Thielemans K, Tsoumpas C, Mustafovic S, Beisel T, Anguiar P, Dikaios N and Jacobson M W 2012 *Phys. Med. Biol.* **57** 867
[19] Siddon R L 1984 *Med. Phys.* **12** 252
[20] Surti S and Karp J S 2008 *Phys. Med. Biol.* **53** 2911

Iterative reconstruction of simulated low count data: a comparison of post-filtering versus regularised OSEM

K Karaoglanis[1], N Efthimiou[2], C Tsoumpas[1,3]
[1] Department of Biomedical Engineering, Division of Imaging Sciences and Biomedical Engineering, King's College London, King's Health Partners, St. Thomas' Hospital, London, SE1 7EH, United Kingdom.
[2] Department of Biomedical Engineering, Technological Educational Institute of Athens, Ag. Spyridonos 28, Egaleo, 12210, Athens, Greece.
[3] Division of Biomedical Imaging, Faculty of Medicine and Health, 8.001a, Worsley Building, Clarendon way, Leeds, LS2 9NL, UK.

e-mail: efthimioun@upatras.gr

Abstract. Low count PET data is a challenge for medical image reconstruction. The statistics of a dataset is a key factor of the quality of the reconstructed images. Reconstruction algorithms which would be able to compensate for low count datasets could provide the means to reduce the patient injected doses and/or reduce the scan times. It has been shown that the use of priors improve the image quality in low count conditions. In this study we compared regularised versus post-filtered OSEM for their performance on challenging simulated low count datasets. Initial visual comparison demonstrated that both algorithms improve the image quality, although the use of regularization does not introduce the undesired blurring as post-filtering.

1. Introduction

Positron emission tomography (PET) is a powerful functional imaging technique that is nowadays widely applied in human clinical practice, primarily in the field of oncology, as well as neurology and cardiology [1].

The image quality of the reconstructed PET images strongly depends on the number of recorded events. Poor statistics lead to images with high noise characteristics and low diagnostic value. Great efforts take place to reduce the injected doses to the patients and the acquisition time. Reduced doses would expose the patients as well as the clinical stuff to lower radioactivity and shorter acquisition times could increase the patients' throughput, giving the opportunity to busy centres to perform more scans, every day. In addition, short scans are less affected by motion artefacts and may even boost the use of radiotracers with shorter half-life [2-4].

The aim of this study is to demonstrate the advantages of Ordered Subsets – Median Root Prior – One Step Late (OS-MRP-OSL) reconstruction algorithm over post-filtered Ordered Subsets – Estimation Maximization (OSEM).

We used realistic PET data of a thorax phantom, analytically simulated from MR data, in which we introduced lesions corresponding to tumours with different intensities and sizes. The study is based on the assumption that shorter than normal scan times and/or reduced injected radioactivity can be represented with increased introduction of Poisson noise.

2. Materials and Methods

2.1. Simulation procedure

MR data from a 1.5 T Philips Achieva^{TM}, using an ultra-short time-echo sequence (UTE) sequence, was obtained. The 3D MR images were segmented and used to produce realistic emission and attenuation data for a Philips Gemini PET. This methodology has been described in detail by Tsoumpas et al [4-7]. For the purpose of this study a dataset with 4.68×10^6 was produced by introducing Poison noise in the emission data of a typical 65×10^6 counts ^{18}F-FDG PET scan.

Spherical lesions corresponding to tumors were analytically inserted in the phantom, in the same manner as Tsoumpas *et al* have demonstrated [4-6]. The lesions' characteristics are listed in Table 1 and the lesions exact location in the phantom is illustrated in Figure 1.

Table 1. Lesion characteristics.

Lesion ID	Diameter (mm)	SUV	Surrounding Tissue
2	14	7.5	Diaphragm
3	14	7.5	Lung
5	14	4.5	Diaphragm
6	14	4.5	Lung
8	8	4.5	Diaphragm
9	8	4.5	Lung

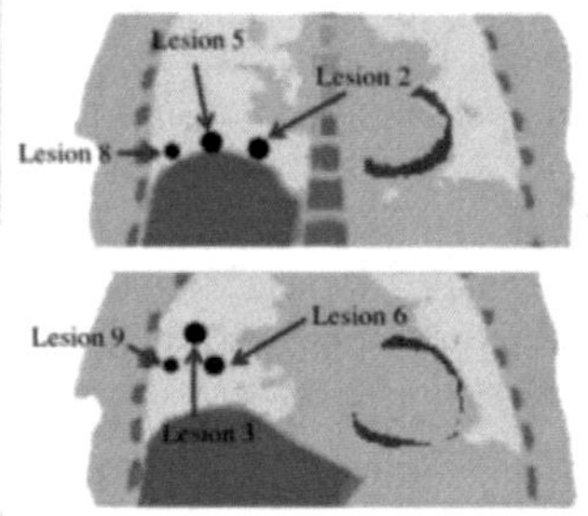

Figure 1. Graphical illustration of the that was used in the simulation procedure and the lesions which were analytically introduced.

2.2. Reconstruction

The OS-MRP-OSL algorithm was originally presented by Green [8,9] and later was implemented in STIR (Software for Tomographic Image Reconstruction) [10], an open source library reconstruction of PET and SPECT images, (**http://stir.sourceforge.net**).

The Wilcoxon signed rank test showed that 10 iterations was a reasonable point to stop and evaluate the results.

2.3. Bias calculation

Bias was calculated for each lesion. Bias calculated by the following formula:

$$\text{bias}_j = \text{mean}_{\text{orig}j} - \text{mean}_j \qquad (1)$$

Where mean_j is the mean of a region *j*, inside a lesion, in the reconstructed images and $\text{mean}_{\text{orig}j}$ is the mean of the same region in the noiseless original images.

3. Results

The coronal view of the region of thorax (lesions 2, 5 and 8) reconstructed with post-filtered OSEM after 10 iterations, is demonstrated in Figure 2. Three cases were considered. No filter (standard

OSEM), 3×3×3 and 6×6×6 mm^3 Gaussian filters, which were applied to the result of each iteration. The application of the filters improved the image quality in the expense of blurring introduction.

Bias was increased for lesions as the FWHM of the Gaussian filter increased. Lesion 2 without filter had bias 3.7, with 3×3×3 mm^3 3.9 and with 6×6×6 mm^3 increased to 4.8. Lesion 5, for the same cases gave values of 1.77, 1.9 and 2.4, while the corresponding values for lesion 1.9 were 1.95, 2 and 2.6.

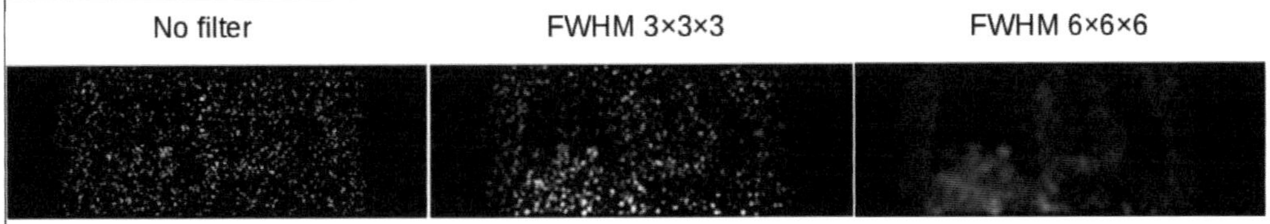

Figure 2. Coronal view of the region of thorax (lesions 2, 5 and 8) reconstructed with post-filtered OSEM.

The coronal view of the region of thorax (lesions 3, 6 and 9) reconstructed with OS-MRP-OSL with different penalization factors (β), after 10 iterations, is demonstrated in Figure 3. The MRP kernel size was set to 3×3×3 to maintain high detail level. Bettinardi et al [11] have shown that details with half the size of the median filter kernel are preserved. Three β values were employed, 0 (standard OSEM), 50 and 75.

By deploying the aforementioned values, the trade-offs for different count levels are to be observed. As one may see the algorithm reduces the noise level without blurring. The optimum value of β depends on the SUV and size of the lesion.

Bias was increased for lesions for β>50. Lesion 3 without regularisation had bias=3.1, with β = 50 was 3.05 and when β = 75 it increased to 3.2. Lesion 6, for the same cases gave values of 2.25, 1.97 and 2.04, while the corresponding values for lesion 9 were 1.95, 1.85 and 2.1.

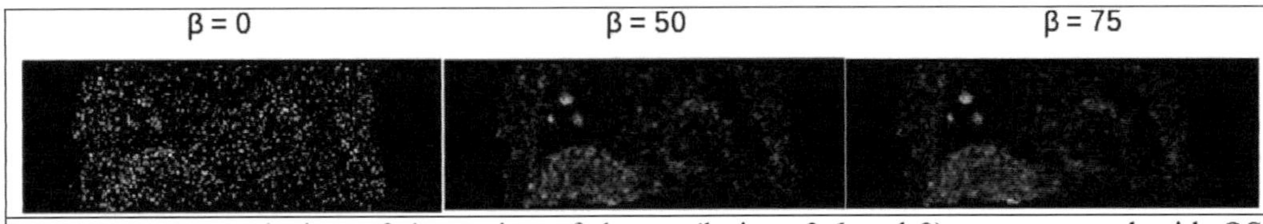

Figure 3. Coronal view of the region of thorax (lesions 3,6 and 9) reconstructed with OS-MRP-OSL with different penalization factors (β), after 10 iterations

Bias calculations

4. Conclusion and Discussion

Summarizing, visual comparison between OS-MRP-OSL and post-filtered OSEM provided indications that under proper configuration both can improve the diagnostic value of low count data. In addition, OS-MRP-OSL does not introduce the undesired effect of blurring. Furthermore, post-filtering increased the bias in the reconstructed images for every lesion, while regularization with β=50 in most cases, reduced the bias. Whether the effect of β will be positive of negative depends on the size and SUV of the lesion.

The selection of the regularization parameter is crucial to the image quality but there is not available any robust, practical automatic way to set this parameter, which among others it depends on the number of measured coincidences.

The evaluation of other priors (e.g. quadratic, anatomical priors) and reconstruction algorithms (e.g. AB-OSEM, NEG-ML) may unveil new prospects.

Acknowledgements
The authors would like to thank A. Gaitanis and G. Spurou (BRFAA, Greece) for the suggestions and fruitful discussions and the IT staff at KCL (Andrew Cantell and Davide Poccecai) for their excellent support. The authors declare no conflict of interest.
The author(s) would like to acknowledge the contribution of the COST Action TD1007.

References
[1] Silva-Rodriguez J, Anguiar P, Dominguez-Padro I, Fierro P, Ruibar A, 2014, Rev Esp Med Nucl Imagen Mol **34**(1), 13-18.
[2] Verdun F R, 2014, Phys Medica **30** : e1
[3] Poon J K, Dahlbom M L, Moses W W et al, 2012,. Phys Med Biol **57**, 4077
[4] Karaoglanis K, Polycarpou I, Efthimiou N, Tsoumpas C, 2015, Hell J Nucl Med (under publication)
[5] Tsoumpas C, Buerger C, King AP, et al, 2011, Phys Med Biol. **56** (20), 6597-613
[6] Tsoumpas C, Gaitanis A., 2013, PET Clin. **8**, 95-110
[7] Buerger C, Tsoumpas C, Aitken A, et al., 2012, IEEE Trans Nucl Sci. **59**(5), 1967-76
[8] Green P J, 1990, IEEE Trans Med Imag. **9**(1), 84-93
[9] Green P J, 1990, J R Stat Soc. **52**, 443-452
[10] Thielemans K, Tsoumpas C, Mustafovic S, Beisel T, Aguiar P, Dikaios N, et al , 2012, Phys Med Biol. **57**(4), 867-83
[11] Bettinardi V, Pagani E, Gilardi MC, et al., 2002, Eur J Nucl Med Mol Imaging, **29**(1), 7-18

Component based normalization method for rotating dual head PET scanner

N Efthimiou[1,2], G Loudos[1], G S Panayiotakis[2]
[1] Department of Biomedical Engineering, Technological Educational Institute of Athens, Greece
[2] Department of Medical Physics, School of Medicine, University of Patras, Greece

E-mail: efthimioun@upatras.gr

Abstract. Component based normalization is a well-established method to calculate correction factors for unbiased and reliable PET reconstruction. Several methods have been studied and validated for cylindrical PET scanners. In this work we adapted a method already presented for cylindrical scanners to rotating dual head PET scanners. The model included corrections for detector efficiency, axial and transaxial geometric effects, crystal interference and attenuation corrections. Results from a simulated realistic dual head PET showed that the adaptation is valid. The images are significantly improved in terms of homogeneity, resolution and background contribution.

1. Introduction

Reliable image reconstruction in Positron Emission Tomography (PET) is only achieved after proper correction of the sensitivity of different lines of response (LOR). Sources of these differences rely in the variations in detector efficiency, geometric effects, mashing of neighbouring LORs etc [1,2].

In direct calculation of the normalization factors, a linear, planar or cylindrical source is used to measure the efficiency of each LOR [1-5]. The main drawback of direct normalization is the need for long acquisitions. On the other hand, component (or model) based normalization, by decomposing the correction factors to several parameters, is possible to accurately estimate the correction factors with relatively shorter acquisitions.

Usually, the parameters are grouped in the detector intrinsic properties and scanner's geometry. In addition, a series of correction factors related to the timing properties of the scanner have been proposed (e.g. dead-time correction, random correction etc) [1-5].

In this study we used a simulated model of a rotating dual head PET scanner, installed in our laboratory. The GATE simulation toolkit was used [6]. The aim was to investigate the application of well validated component based normalization algorithms from cylindrical PET scanners [7,8]. We used a planar source and no rotation in order to calculate the correction factors and extrapolated analytically the resulted factors the over 180° angular coverage. In order to validate our calculations we used two phantoms a uniform cylindrical and an assembly of 21 capillary cylinders evenly spaced every 1 cm.

2. Material and Methods

2.1. GATE simulations

The geometry of a dual head PET scanner, installed in our laboratory, was used as a prototype for the development of the simulation's geometry. The detailed scanner's presentation and evaluation as well as the validation of the simulation models have been presented previously [9,10]. Briefly the scanner is based on two Hammamatsu H8500 PSPMTs which are coupled with two 20×20 LSO:Ce crystal matrices, with crystal size 2×2mm^2.

The two heads were rotated in unison for 180°, with 10° step and the acquisition duration on each step was 60 sec. The head separation distance was 5 cm and the ^{176}Lu intrinsic radioactivity was taken into consideration. The low level threshold was set to 200 keV and the energy window was set to 350-650 keV.

The simulated data were bined into 3D sinograms, with bin size ~0.3 mm. The angle of rotation and the angle of the detector with respect to the center of the head's rotation were summed for the calculation of azimuthial angle of each LOR, as illustrated in Figure 1(a). The 88 angular positions reflect the sum of events over all detectors located every ~2° during in each step of rotation, as illustrated in Figure 1(b).

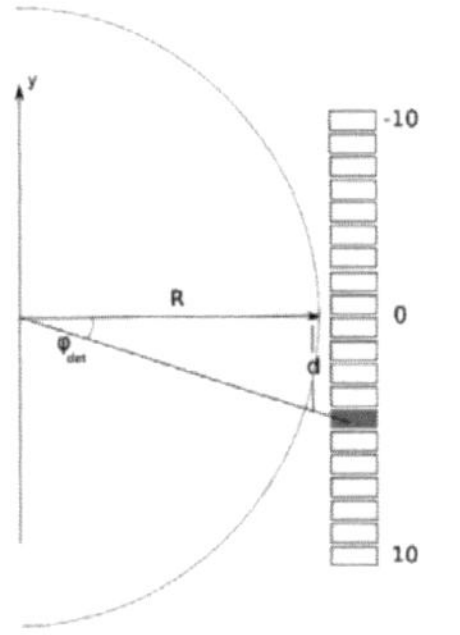

(a)

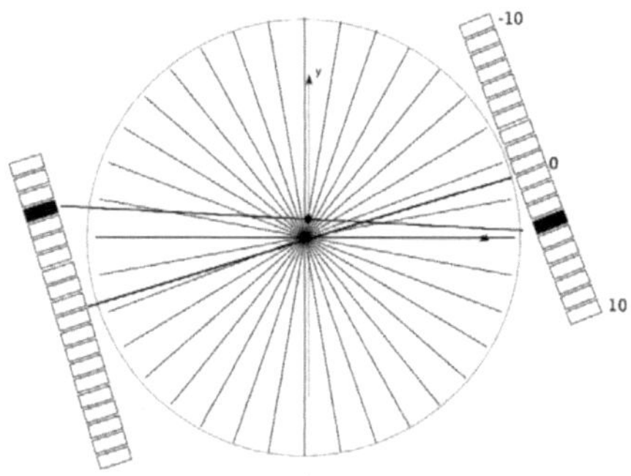

(b)

Figure 1. (a) For calculation of the azimuthal angle of each event the angle of the detector with respect to the centre of the head was calculated. (b) The azimuthal angle of the detector was added to the rotation angle of the heads for the calculation of the LOR's azimuthal angle.

2.2. Phantoms

In this study two phantoms were used. First a homogeneous water cylindrical phantom (Ø 4cm) coupled with a ^{68}Ga ion source. Second a phantom assembled by 21 cylindrical tubes made from water, evenly spaced every 1cm. The diameter of each capillary was 1.1 mm in order to simulate the size of regular capillary tubes.

The data were reconstructed using a full 3D MLEM algorithm implemented in STIR [11] using 5 subsets after 5 subiterations. The mean depth of interaction inside the crystal was taken into account. But no arc-correction, random, scattered and dead-time corrections were performed.

2.3. Normalization components

For the calculation of the normalization components a 7×7 cm^2 planar phantom located in the centre of the FOV, was simulated only for one rotation step. The simulated data then were extrapolated using geometrical calculations for all rotation steps. The low hardware threshold was set high to 200 keV with the energy window from 350-750 keV. This configuration was chosen in order to minimize the contribution of low energy photons to system's dead-time while. In addition, it has been shown that in small animal imaging the contribution of scattered photons down to 350 keV is not significant, while the system benefit from the higher count rate [12].

The normalization components that were included in the model were the axial and transaxial geometric factor, block sensitivity correction, crystal interference, detector efficiency and attenuation correction. Even though in GATE detectors have uniform efficiency it has been shown that a correction should be applied [8].

Journal of Physics: Conference Series **637** (2015) 012007 doi:10.1088/1742-6596/637/1/012007

3. Results

The comparison between the non-normalised and normalized images from the cylindrical and the 21 capillaries phantom, are displayed in Figure 2(a-d). In Figure 2(e-h), one may see their corresponding profiles, which represent an average of the three central rows. As one can see in Figures 2(a) and (b) the homogeneity inside the cylinder is strongly improved. Near the edge of the cylinder is observed a depth of iteration artefact. Furthermore, a strong contribution of scattered photon is observed, the use of STIR's scatter correction function should help reduce this effect [13].

Regarding the 21 capillaries phantom, before the normalization the peaks for each capillary are higher toward the centre of the FOV and a strong background area is located around the centre. After normalization we get similar intensity values for each capillary, the background in the centre is strongly reduced and a resolution recovery is observed. The shape of the capillary's profile towards the edges of the FOV is gradually deformed, which is a common effect due to the parallax effect.

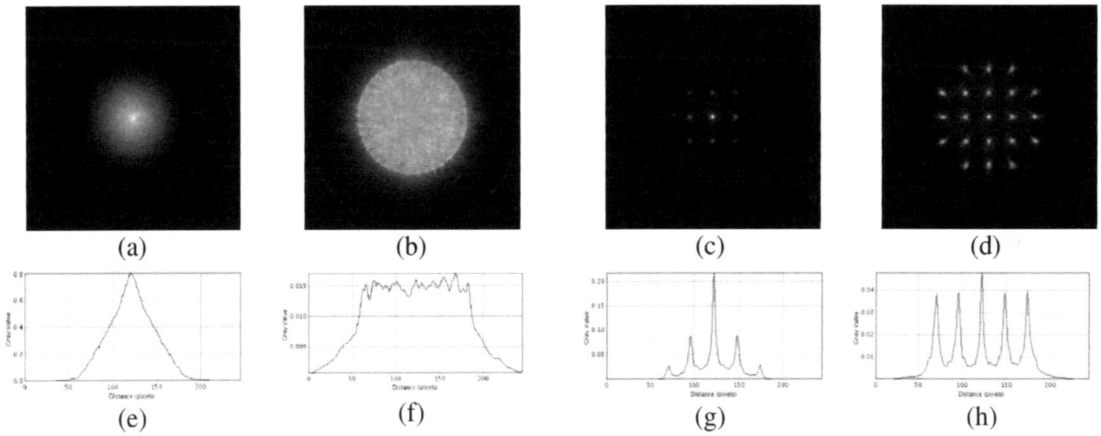

Figure 2. Non-normalised image from the cylindrical (a-b) and 21 capillaries phantom (c-d) and their corresponding profiles (e-f) and (g-h). The profiles are the average of the three central rows.

4. Conclusion and Discussion

Normalization for PET data from a dual head planar PET may be performed using directly adapted algorithms from cylindrical PET scanners. The acquisition of the normalization data can be performed only for one acquisition step using a planar source and then use geometric calculations to calculate the correction factors for the rest of the rotation steps..

Acquisition of data for one rotation angle significantly reduces the normalization scan time but is susceptive to errors like error of rotation and errors in the placement of the phantom.

By correcting for detector efficiency, axial and transaxial geometric effects, block sensitivity and attenuation the reconstructed images are strongly improved near the centre, in terms of homogeneity, intensity and background noise. Introduction of more correction components in our model will further improve the quality of the images on the expense of additional complexity of the algorithm.

Acknowledgements

This research has been co-financed by the European Union (European Social Fund-ESF) and Greek national funds through the Operational Program "Education and Lifelong Learning" of the National Strategic Reference Framework (NSRF)-Research Funding Program: Heracleitus II.

This research has been co-funded by the European Union (European Social Fund) and Greek national resources under the framework of the "ARISTEIA II" project HYPERGNOSTIC code 4309 of the "Education \& Lifelong Learning" Operational Programme.

References

[1] Bailey L D, Townsend W D, Kinhan E P, Grootoonk S, Jones T, 1996, IEEE Trans. on Nucl. Sci. **43**, 3300-07

[2] Badawi R D, Ferreira N C, Kohlmyer S G, Dahlbom M, Marsden P K, Lewellen T K, 2000, Phys. Med. Biol. **45**, 3253-66.

[3] Bai B, Li Q, Holdsworth C H, Asma E, Tai Y C, Chatziioannou A, Leahy R M, 2002, Phys. Med. Biol. **47**, 2773-84.

[4] Rodriguez M, Liow J S, Thada S, Sibomama M, Chelikani S, Mulnix T, Johnson C A, Michel C, Craig Barker W, Carson R E, 2007, IEEE Trans. on Nucl. Sci. **54**, 486-95

[5] Wang W, Hu Zhiqiang, Gagnon D, 2007, IEEE Trans. on Nucl. **54**, 92-9

[6] Sarrut D, Bardiès M, Boussion N, Freud N, Jan S, Létang JM, Loudos G, Maigne L, Marcatili S, Mauxion T, Papadimitroulas P, Perrot Y, Pietrzyk U, Robert C, Schaart D, Visvikis D, Buvat I, 2014, Med. Phys. **41** :064301.

[7] Pepin A, Stute S, Jan S, Comtat C, Normalization of Monte Carlo PET data using GATE, 2011, IEEE Nucl Sci Conf R 4196- 4200

[8] Vicente E, Vaquero J J, Espana S, Herraiz J L, Udias J M, Desco M, 2006, IEEE Nucl Sci Conf R, M06-379

[9] Efthimiou N, Maistros S, Tripolitis X, Samartzis A, Loudos G, Panayiotakis G S, 2011, Meas. Sci. Technol. **22**, 114010

[10] Efthimiou N, Loudos G, Karakatsanis N A, Panayiotakis G S, 2013, Phys. Med. **40**, 112505

[11] Thielemans K, Tsoumpas C, Mustafovic S, Beisel T, Aguiar P, Dikaios N, and Jacobson M W, 2012, Phys. Med. Biol. **57** (4), 867-83.

[12] Efthimiou N, Loudos G, Karakatsanis N, Panayiotakis G S, 2013, IEEE Trans. on Nucl. Sci., pp 1-3.

[13] Polycarpou I, Thielemans K, Manjeshwar R, Anguiar P, Marsden P K, Tsoumpas C, 2011, Ann. Nucl. Med. **25**, 643-9.

Optimal Elasticity cut-off value for discriminating Healthy to Pathological Fibrotic patients employing Fuzzy C-Means automatic segmentation in Liver Shear Wave Elastography images

Ilias Gatos[1], Stavros Tsantis[1], Aikaterini Skouroliakou[2], Ioannis Theotokas[3], Pavlos S. Zoumpoulis[3] and George C. Kagadis[1]

1. Department of Medical Physics, School of Medicine, University of Patras, GR 26504, Rion
2. Department of Energy Technology Engineering, Technological Educational Institute of Athens, GR 12210, Egaleo
3. Diagnostic Echotomography SA, 317C Kifissias Ave., GR 14561, Kifissia

E-mail: gatose1981@gmail.com

Abstract. The aim of the present study is to determine an optimal elasticity cut-off value for discriminating Healthy from Pathological fibrotic patients by means of Fuzzy C-Means automatic segmentation and maximum participation cluster mean value employment in Shear Wave Elastography (SWE) images. The clinical dataset comprised 32 subjects (16 Healthy and 16 histological or Fibroscan verified Chronic Liver Disease). An experienced Radiologist performed SWE measurement placing a region of interest (ROI) on each subject's right liver lobe providing a SWE image for each patient. Subsequently Fuzzy C-Means clustering was performed on every SWE image utilizing 5 clusters. Mean Stiffness value and pixels number of each cluster were calculated. The mean stiffness value feature of the cluster with maximum pixels number was then fed as input for ROC analysis. The selected Mean Stiffness value feature an Area Under the Curve (AUC) of 0.8633 with Optimum Cut-off value of 7.5 kPa with sensitivity and specificity values of 0.8438 and 0.875 and balanced accuracy of 0.8594. Examiner's classification measurements exhibited sensitivity, specificity and balanced accuracy value of 0.8125 with 7.1 kPa cutoff value. A new promising automatic algorithm was implemented with more objective criteria of defining optimum elasticity cut-off values for discriminating fibrosis stages for SWE. More subjects are needed in order to define if this algorithm is an objective tool to outperform manual ROI selection.

1. Introduction

Chronic Liver Disease (CLD) is considered as the only major cause of death still increasing year-on-year, causing more deaths than diabetes and road accidents combined [1].For this reason the correct estimation of the disease progress towards a suitable treatment is considered very important.

Liver Biopsy (LB) plays an important part of patient evaluation to various liver diseases. Besides establishing the diagnosis, the biopsy is often used to assess the severity of the disease in terms of both grade and stage. Staging in most chronic liver diseases relates to the degree of scarring (fibrosis) with cirrhosis with its clinical complications being the final stage [2]. Although LB is still considered as 'gold standard' for estimating the progress of CLDs it has many important limitations. It is invasive and can cause significant complications. Nearly 30% of patients report having substantial pain after liver biopsy, and some experience serious complications such as pneumothorax, bleeding, or puncture of the biliary tree. In rare cases, patients die of bleeding. Furthermore, hepatic pathology, particularly fibrosis, is not always uniformly distributed. Surgical wedge biopsy provides adequate tissue volume to overcome this problem.

Needle biopsy, on the other hand, provides a much smaller volume of tissue (1/50,000 of the total liver mass) [3]. Obviously, needle liver biopsy is far from an ideal test. Liver diseases can be diagnosed precisely with laboratory tests, imaging studies, or both. Thus, needle liver biopsy is playing a lesser role in diagnosis.

Conventional imaging techniques cannot provide information about tissue mechanical properties. Many injuries can cause changes in tissue stiffness, especially tumors and fibrosis. In recent years, various non-invasive ultrasound methods have been developed to study tissue elasticity for a large number of applications [4].

The ultrasound based elastographic methods used for liver fibrosis assessment can be divided into two main categories: strain elastography (or quasi-static elastography) – in which tissue excitation is performed manually or using the cardiac motion; and shear wave elastography – in which tissue excitation is induced either by ultrasound waves produced by the probe (for elastographic methods integrated into a standard ultrasound system: Acoustic Radiation Force Impulse Elastography and Real Time 2D-Shear Waves Elastography), or induced by a mechanical impulse (thump) generated by a coaxial vibrator (for Transient Elastography).

Shear wave elastographic methods include: Transient Elastography (TE), Acoustic Radiation Force Impulse (ARFI) quantification Elastography and Real Time 2D-Shear Waves Elastography (2D-SWE) [5].

TE is the oldest shear wave elastographic method, used for liver fibrosis evaluation in various liver diseases. The usefulness of TE for liver fibrosis assessment was confirmed by meta-analyses studies. ARFI elastography was more recently developed with data published in the last years showing that it is a useful tool for fibrosis evaluation, non-inferior to TE. SuperSonic Shear Imaging (2D-SWE) is the latest developed shear waves elastographic method with few but promising data being available until now [5-6].

Most approaches on literature try to set cut-off values to correspond elasticity values to fibrosis stages according to Metavir Classification. The clinical protocol used depends on taking the mean number value of successive liver measurements of tissue elasticity from the same intercostal space. ROC analysis is performed then on all the mean elasticity values from all patients to obtain the best cut-off values which delimit Elasticity values ranges that correspond to Metavir Fibrosis Stages [5-6].

This approach shows relatively good results for the best cut-off values (0.8-0.85 AUC) but has quite serious limitations. The examiner places a Region of Interest (ROI) superimposed on B-Mode Ultrasound on a suitable area of liver parenchyma which shows the stiffness of the liver by applying elasticity values to a color map (low stiffness: blue, high stiffness: red) so that there is a color specification for liver stiffness for the examiner to take successfully the measurement. This measurement is taken by placing a circular shaped index in the ROI to take the mean stiffness of the area inside the circular index [5-6].

This method relies on the examiner's opinion of where is the most representative area of liver parenchyma to take the measurement. This means that the method is subjective to the examiner's criteria who in most cases uses prior knowledge of patient's clinical history, blood serum markers, or histological examination. For example a patient with a clinical history of cirrhosis provokes the examiner to choose stiffer values as most representative. This also leads to significant Inter- and Intra-observer variability.

The other issue about this method is that takes only two features of the histogram of the elasticity values (Mean and Standard Deviation) and doesn't take advantage of the rest information that lies in the image. As even an experienced Radiologist cannot notice every detail or factor on the image that

could assist in diagnosis, the role of an automatic system that can calculate the most significant features of an image for diagnosis is very important.
Only one automated approach towards this direction has been done [7]. The study's main purpose was to estimate how factors such as stability of an area in time and low heterogenicity define areas that their mean values can improve the reliability and reproducibility of SWE using the proposed method.
Our study tries to define factors that are important to estimate fibrosis stage from elasticity measurements and make a semi-automated system that prognoses accurately the corresponding fibrosis stages. This is done by segmenting the elastographic image to clusters with low heterogeneity and choosing the cluster with maximum number of pixels. Then the mean stiffness value of the chosen cluster is calculated and is used for ROC analysis to define best cut-off values.
As our data consists of few patients preliminary results are presented and only healthy subjects with no fibrosis (F0) are discriminated from patients with CLD (F1, F2, F3, F4).

2. Materials & Methods

2.1. Clinical Data
Our clinical material consisted of 32 subjects (16 healthy (F0) and 16 with CLD (F1, F2, F3, F4)) who were examined by an experienced radiologist. All SWE measurements were performed in the right lobe of the liver using an appropriate intercostal space. The distance probe / ROI was 3-6 cm. All SWE measurements were performed on a homogenous B-Mode Ultrasound image with no Ultrasound visible vessels. A normal inspiration was established for diaphragm stabilization assuring liver's immobility. When an imperfect skin/probe contact was observed, an acoustic coupling medium (pad) was interposed to assure a perfect acoustic transmition.
When a SWE image could not be obtained using an intercostal space, the left lobe of the liver was used for SWE measurements. In both circumstances perfect immobility of the liver parenchyma and minimum disturbance from heart's compression were assured.
Two SWE images for each subject were saved, one with no markings on the elastogram for testing the proposed method and one with the radiologist's stiffness measurement using the standard protocol that is provided by the SSI Aixplorer and that is providing the mean value of a circular ROI placed on the elastogram Q-Box superimposed on B-Mode Ultrasound image.
All healthy subjects had normal Biochemical Markers and with no abnormal liver findings during their Ultrasound examination. All CLD patients performed liver biopsy and histological examination for fibrosis stage estimation by a senior histopathologist for validation purposes.

2.2. Automatic ROI Detection
SWE elastogram images with no markings were segmented by means of Fuzzy C-Means (FCM) unsupervised clustering, utilizing 5 clusters for each image. Mean Stiffness value and pixels number of each cluster were calculated using Aixplorer's colormap for inverting RGB values to stiffness values. The mean stiffness value feature of the cluster with maximum pixels number was then fed as input for ROC analysis.

3. Results
The selected Mean Stiffness value feature an AUC of 0.8633 (Figure 1) with optimum Cut-off value of 7.5 kPa for F≥F1 with Sensitivity and Specificity values of 0.8438 and 0.875 and Balanced Accuracy of 0.8594. Automatic ROIs sizes defined by the proposed algorithm ranged from 0.5-0.8 of the whole elastogram.
Examiner's classification measurements exhibited Sensitivity, Specificity and Balanced Accuracy values of 0.8125 with 7.1 kPa cut-off value.

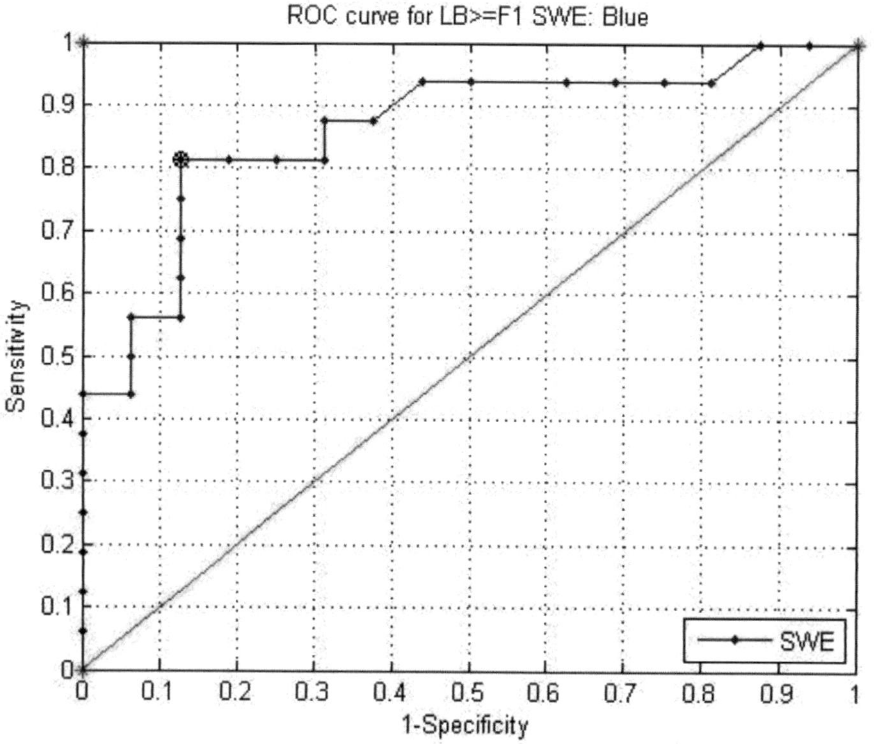

Figure 1. ROC Curve for Mean Stiffness value feature.

4. Discussion

A new promising automatic algorithm was implemented with more objective criteria of defining optimum elasticity cut-off values for discriminating fibrosis stages for SWE. The objectivity of the proposed method is achieved by defining standard criteria (maximum size homogeneous area) that discriminate more accurately Metavir fibrosis stages than the existing methods. This could lead to a software that selects and takes measurement automatically from a suitable area or propose an area for the examiner to take measurement with the existing circular ROI. As a next step for research purposes, a study with two radiologists taking two measurements each, by the same proposed area could evaluate if the proposed method improves intra- and inter-observer variability achieving better accuracy from the existing method. More subjects are needed in order to discriminate between all fibrosis stages and to define if this algorithm is an objective tool to outperform manual ROI selection.

5. References

[1] Stevenson M, Lioyd-Jones M, Morgan MY and Wong R 2012 *Health Technol. Assess.* **16** 1-176
[2] Goodman ZD 2007 *J. Hepatol.* **47** 598-607
[3] Carey E and Carey W 2010 *Cleve. Clin. J. Med.* **77** 519-27
[4] Frulio N and Trillaud H 2013 *Diagn. Interv. Imaging* **94** 515-34
[5] Sporea I, Bota S, Gradinaru-Tascau O, Sirli R, Popescu A and Jurchis A 2014 *Eur. J. Radiol.* **83** 118-22
[6] Deffieux T *et al.* 2015 *J. Hepatol.* **62** 317–24
[7] Pellot-Barakat C, Lefort M, Chami L, Labit M, Frouin F, and Lucidarme O 2015 *Ultrasound Med. Biol.* **41** 936-43

Journal of Physics: Conference Series **637** (2015) 012009 doi:10.1088/1742-6596/637/1/012009

The electromagnetic environment of Magnetic Resonance Imaging systems. Occupational exposure assessment reveals RF harmonics.

G Gourzoulidis[1,3], E Karabetsos[2], N Skamnakis[2], C Kappas[3], K Theodorou[3], I Tsougos[3] and T G Maris[4]
[1]Hazardous Agents Dep., Hellenic Ministry of Labor, 29 Stadiou str., GR10110
[2]Greek Atomic Energy Commission, Non-Ionizing Rad. Off., POBox60092, GR15310
[3]Department of Medical Physics, University of Thessaly, GR41500, Larissa
[4]Department of Medical Physics, University of Crete, POBox1352, GR71201

E-mail: ggourz@yahoo.com

Abstract. Magnetic Resonance Imaging (MRI) systems played a crucial role in the postponement of the former occupational electromagnetic fields (EMF) European Directive (2004/40/EC) and in the formation of the latest exposure limits adopted in the new one (2013/35/EU). Moreover, the complex MRI environment will be finally excluded from the implementation of the new occupational limits, leading to an increased demand for Occupational Health and Safety (OHS) surveillance. The gradient function of MRI systems and the application of the RF excitation frequency result in low and high frequency exposures, respectively. This electromagnetic field exposure, in combination with the increased static magnetic field exposure, makes the MRI environment a unique case of combined EMF exposure. The electromagnetic field levels in close proximity of different MRI systems have been assessed at various frequencies. Quality Assurance (QA) & safety issues were also faced. Preliminary results show initial compliance with the forthcoming limits in each different frequency band, but also revealed peculiar RF harmonic components, of no safety concern, to the whole range detected (20-1000MHz). Further work is needed in order to clarify their origin and characteristics.

1. Introduction

A few years ago, European countries put into force legislative acts to implement the 1999 Recommendation of the Council of the European Union for the protection of the general public from Electromagnetic Fields (EMF) exposure. This Recommendation, based on ICNIRP's (International Commission for Non-Ionizing Radiation Protection) guidelines for EMF (0-300GHz) [1], also formed the basis for the 2004/40/EU Directive concerning the protection of workers from EMF exposure, but this Directive was never implemented due to the strict static magnetic field occupational limits stated, that would restrict the use of MRI systems. ICNIRP made significant changes in its guidelines for the 0-100kHz frequency region in 2010 [2]. These new guidelines, together with the former ones, which are still valid for high frequencies, formed the basis for the new Directive (2013/35/EU) [3]. This is part of the overall legislation for Occupational Health and Safety (OHS) and during its implementing process into national law, various work places (including MRI [4]) are assessed for EMF occupational exposure, through the collaboration of the Ministry of Labor with the Non-Ionizing Radiation Office of the Greek Atomic Energy Commission (GAEC). Note that the complicated Quality Assurance (QA) and safety issues in MRI are covered in Greece by a complete QA MRI protocol [5].

2. Materials and methods

2.1. NMR and MRI electromagnetic characteristics

Magnetic Resonance Imaging (MRI) is a term invented by radiologists to describe Nuclear Magnetic Resonance (NMR) imaging, a spectroscopic method originally concerned with identifying signals from chemical compounds using small probes (cylinders: 0.2-0.5cm diameter) with no spatial information. Since the 70's, NMR imaging has evolved as a technique where signal spatial localization with the aid of magnetic gradient fields was feasible. Signals were then obtained from larger probes (cylinders: 1–3cm). Finally, since the 80's and based on the same localization principles, MRI was introduced as a standard diagnostic method probing the entire human anatomy. As expected, in clinical MRI, signals were then obtained from extremely large probes (cylinders: 60-70cm). In both NMR and MRI techniques, the final signal is detected from selected nuclei being present and magnetically coupled in atoms/molecules, which in turn are moving in different magnetic molecular environments. These magnetic couplings and the actual physical atomic/molecular movements modulate the final signal, which is strongly related to either the magnetic parameters, known as T1, T2 and T2*, or the physical parameters, like molecular diffusion or both. The most studied nucleus in biological NMR and in clinical MRI is the nucleus of the Hydrogen atom. More specifically, NMR and MRI are based on the magnetic excitation of nuclei being present inside a homogeneous, high static magnetic field, ranging from 0.2 up to 40T. In clinical MRI, static fields range from 0.2 up to 8T. Nowadays, NMR and MRI systems are implementations of pulsed nuclear excitation/detection experiments. This means that in both techniques, nuclei are excited utilizing bursts of intense radiofrequency pulses. These pulses are centered on a basic excitation frequency, which is linearly related to the static field strength and in the clinical MRI lies in the range of 8MHz (0.2T) up to 340MHz (8T).

In either NMR or MRI systems the final signal is detected just after the excitation RF pulse, as the nuclei return to thermal equilibrium by emitting time domain recordable RF signals, with basic frequencies lying on the MHz spectrum and depending on the external static field strength. This relaxation procedure is controlled by specific relaxation time constants, namely T1, T2, T2* and the decaying signal contains the sum of the frequencies from all the target nuclei. Most of this signal is picked up in the coil, as an oscillating EMF, generated by the nuclear magnetic moments rotating back to equilibrium. The actual signal cannot be recorded directly, because its frequency is too high (MHz range). It is mixed with a lower frequency carrier signal to produce a low frequency interferogram, which is further digitized and called the Free Induction Decay (FID) signal. Fourier transformation of the FID yields a frequency domain spectrum, which is the final recorded signal [6]. It is the authors' assumption that nuclei relaxation is not a single dominant frequency phenomenon and their harmonic components pattern should be strongly related to the nuclei magnetic environment [4].

In clinical MRI, spatial localization is achieved by the superimposition of powerful linear magnetic gradient fields ranging from 10 up to 80mT/m and operating in a switching mode with slew rates of 50 up to 300mT/m/ms. Gradient fields are used for slice selection during signal excitation and signal detection steps, frequency encoding and phase encoding procedures during digital imaging reconstruction steps. Gradient switching is achieved by transmitting signals on the kHz range [7].

2.2 The MRI in the new EMF Directive (2013/35/EU)

The new EMF Directive implies a system of multiple exposure limits, with the two basic definitions being: 'Exposure limit values (ELVs)' and 'Action levels (ALs)' [3]. ELVs are established on the basis of scientifically well-established, short-term and acute direct effects and refer to the internal field (into the human body). They are the substantial limits in order to check for health consequences and there is no way to have direct measured values compared to them, but only through complex computer modeling [8]. ALs are operational levels established for the purpose of simplifying the process of demonstrating the compliance with relevant ELVs, since external field measurements are directly comparable to ALs. AL overexposure does not necessarily mean that ELVs are also exceeded [9].

The Directive excludes, under a prescribed scheme, MRI systems from the application of the ELVs [3], but MRI personnel are covered by OHS legislation even in the case of derogation. On the contrary, patients whose exposure can be much higher are not considered as general public in order to be covered by the corresponding legislation, but the Justification and Optimization principles are applicable. In both cases the implementation of best practices has multiple benefits.

Measurements of RF electric field and ELF magnetic field were performed next to the examination bed, during the MRI scans for 1.5T and 3T systems, during occupational exposure assessment [4]. The use of Narda IDA-3106 Interference and Direction Analyzer revealed low amplitude RF harmonic components. Three different directional antennas were used.

3. Results

Spectrums and spectrograms (spectrum pattern as a function of time) were obtained with the Narda IDA-3106 for 1.5T systems. The RF safety issues are linked with the 63.5MHz frequency, which is clearly visible and around it many small others harmonics, of no safety concern, exist (figure 1).

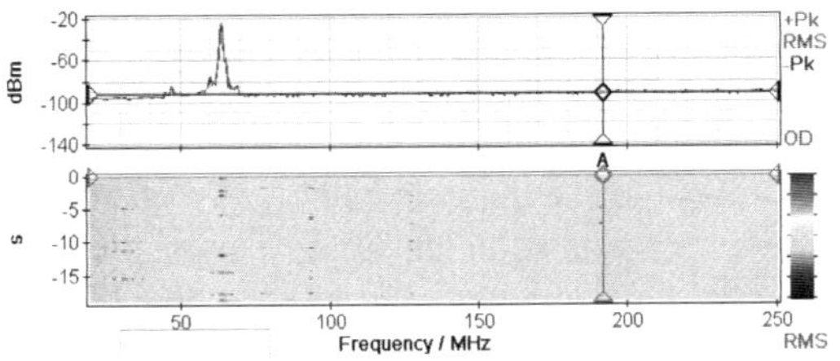

Figure 1. Relative comparison between 63.5MHz and harmonic components for 1.5T.

Harmonics can be seen for the 200-500MHz (figure 2) and for the 400-1000MHz regions.

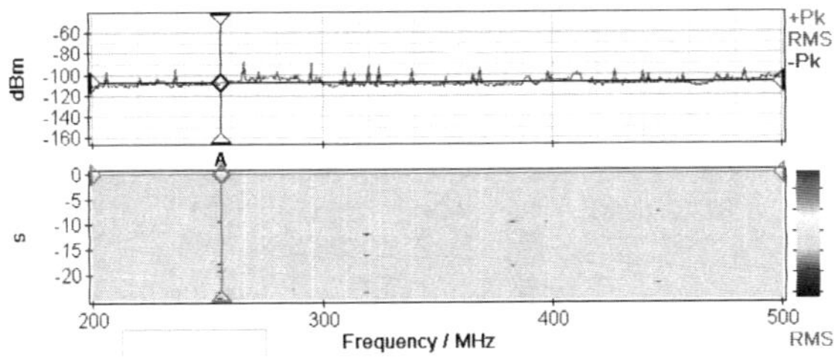

Figure 2. RF harmonic components between 200 and 500 MHz for a 1.5T MRI system.

The RF shielding check with Narda IDA-3106 of one system revealed FM radio interference due to some copper lamina missing from the door (figure 3).

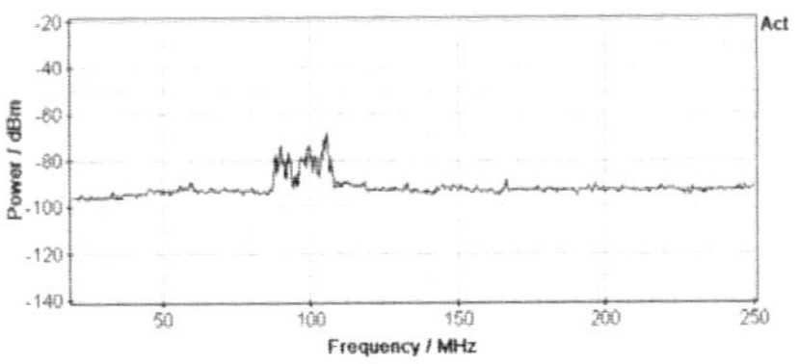

Figure 3. Faulty RF shielding.

4. Discussion

MRI, apart from a very important imaging technique, is also a very interesting and rich electromagnetic environment. Almost the entire EMF spectrum is present during an MRI scan, resulting in a combined exposure to the static & ELF magnetic field and to the RF electromagnetic field. SCENHIR (Scientific Committee on Emerging and Newly Identified Health Risks, of the European Commission) classifies the investigations of suggested health effects in the MRI environment as highly important and urgent [10].

Initial occupational MRI investigation showed compliance with the forthcoming limits, but also revealed peculiar RF harmonic components [4]. The first results presented need to be expanded and clarified for different equipment and clinical practices.

Shielding integrity of the MRI room can be checked by recording the background RF spectrum inside the room and surveying around it. In this way small shielding discontinuities can be identified.

5. Conclusions

The RF harmonic components found are of no safety concern. Their origin is not clear and it is under examination whether they follow any specific pattern or not. Harmonics of the main 63.5MHz frequency are detected, but also other submultiple ones and even the half of the main frequency, are present (overtones and semitones). This seems to be repeated to higher frequencies too. It is the authors' suggestion that the observed RF frequency patterns are somehow related to the magnetic molecular environments, on which the nuclei are bound or interact with. Moreover, the RF frequency patterns were revealed with the use of directional antennas and were also direction dependent. Further research studies on the specific topic have to be meticulously designed. Questions relating the type of the materials (water, fat, gelatin, paramagnetic solutions, etc.), their actual shapes (e.g. filling hollow spaces) and magnetic field strength, with a specific RF pattern have to be answered.

Acknowledgements

We would like to acknowledge: The Director of ACTA Ltd, Mr. A. Georgiou, as well as Mr. N. Marinakis, for being so kind to provide us, free of charge, with the IDA- 3106 Narda equipment.

References

[1] ICNIRP Guidelines for limiting exposure to time-varying electric, magnetic, and electromagnetic fields (up to 300 GHz) 1998 *Health Phys* **74** 4 pp 494-522.

[2] ICNIRP factsheet on the guidelines for limiting exposure to time-varying electric and magnetic fields (1Hz-100kHz) 2010 *Health Phys* **99** 6 pp 818-836.

[3] Directive 2013/35/EU of the European Parliament and of the Council of 26 June 2013(EMF) *Off. Jour. of the European Union* L 179/1.

[4] Gourzoulidis G, Karabetsos E, Skamnakis N, Maris T G et al 2015 Occupational EMF exposure in MRI systems–RF harmonic content *Phys Med: Eur Journal of Med Phys* 10.1016/j.ejmp.2015.03.006.

[5] MRI Quality Assurance protocol, http://eeae.gr/docs/files/_qa_mri.pdf February 2013.

[6] Purcell E M et al 1946 Res. absorption by nuclear magnetic moments in a solid *Phys. Rev.* **69** 37.

[7] MRI e-learning, www.imaios.com 1.12.2014.

[8] Bakker J F et al 2012 Children and adults exposed to low-frequency magnetic fields at the ICNIRP ref. levels: theoretical assessment of the induced electric fields *Phys. Med. Biol.* **57** pp 1815–1829.

[9] Neubauer G et al 2011 Exposure Assessment in Complex Welding Environment *Austrian institute of Technology, Seibersdorf,* http://www.cost-action-bm0704.org/doc/Neubauer-welding.pdf.

[10] SCENIHR Opinion on potential health effects of exposure to EMF 2015 *European Commission* http://ec.europa.eu/health/scientific_committees/emerging/docs/scenihr_o_041.pdf, 20.01.2015.

Journal of Physics: Conference Series **637** (2015) 012010 doi:10.1088/1742-6596/637/1/012010

Development of the RF Front-end of a Multi-Channel Microwave Radiometer for Internal Body Temperature Measurements

Sotiris Karabetsos[1], Grigorios Koulouras[1], Panagiotis Charamis[1], George Adamidis[2], Ioannis O Vardiambasis[2], Athanasios Nassiopoulos[1]
[1]RDETL Laboratory, Department of Electronic Engineering, Technological Educational Institute (TEI) of Athens, Athens, Greece.
[2]Department of Electronic Engineering, Technological Educational Institute (TEI) of Crete, Crete, Greece.

E-mail: sotoskar@teiath.gr, gregkoul@teiath.gr, nassiop@teiath.gr.

Abstract. Microwave Thermograph (MT) is based on measuring the electromagnetic field spontaneously emitted by a body in micro-wave frequency range. In microwave radiometers, temperature measurement is made by measuring the thermal noise power. The primary module that is used for detecting this thermal noise power is the RF Front-end, which must meet very challenging requirements in terms of accuracy for measuring the noise power at the input of the receiver. The work that will be presented here will exhibit the design approaches and specifications as well as the trade-offs and performance criteria towards the development and prototyping of the RF Front-end for a Multi-Channel Microwave Radiometer for internal body temperature measurements in the 1-4GHz frequency bands. The RF Front-end is intended to be integrated and be a part of a full Microwave Radiometer device that targets early detection of malignant tumours. The latter relate with the increase of temperature in cancerous cells and the precise detection of the temperature difference, which is the goal of the radiometer.

1. Introduction

Microwave Thermograph (MT) is based on measuring the electromagnetic field spontaneously emitted by a body in micro-wave frequency range [1]. In microwave radiometers, temperature measurement is made by measuring the thermal noise power. That is, microwave radiometers are instruments for resolving blackbody radiation of heated matter above absolute zero. The emission levels of blackbody radiation vary with frequency. Medical microwave radiometers are non-invasive, non-toxic, absolutely safe, and can be used to form a thermal image from the tissues of the human body [1], [2], [3]. An application field of great interest for the microwave radiometer is the diagnosis of breast cancer. The primary module that is used for detecting this thermal noise power is the RF Front-end, which must meet very challenging requirements in terms of accuracy for measuring the noise power at the input of the receiver (antenna) [3]. The most common architecture of the RF Front-end found in the literature is the so called Dicke radiometer. A typical Dicke RF Front-end consists of an antenna, a noise reference, an electronic switch, a Low Noise Amplifier (LNA), a Band-Pass Filter (BPF) with

[1] Corresponding Authors: Sotiris Karabetsos, sotoskar@teiath.gr, Grigorios Koulouras, gregkoul@teiath.gr

appropriate Q factor, a booster amplifier, as well as additional components such as a square law power detector, a low frequency (LF) amplifier, an Analog to Digital Converter (ADC), and a CPU controller that comprise the full radiometer [3], [4]. Given the principle of practical multi-stage cascaded radiometric systems, the noise performance of the first stage is critical. This inherent property is often utilized in measurement systems through implementation of low-noise preamplifiers close to the sensor element; furthermore, it leads to the postulate that achieving a high sensitivity requires a wide band implementation and large time integration, whereas a narrow band design is balanced by larger time integration [1], [4]. The work that will be presented in this paper, exhibits the design approaches and specifications as well as the trade-offs and performance criteria towards the development and prototyping of the RF Front-end for a Multi-Channel Microwave Radiometer for internal body temperature measurements in the frequency region between 1GHz and 4GHz.

2. Design and Development of the RF Front-end

Figures 1 and 2 depict the architecture of the radiometer together with the investigated RF Front-end. Our approach comprises of a compact (patch) single-element wideband antenna able to receive microwave signals in the frequency region from 1GHz to 4GHz. The antenna interfaces with a Dicke microwave radiometer. In addition, a Band-Pass Filter selection circuit is responsible for the desired operating RF band (Table 1), which corresponds into different penetrating depths inside the human tissue; hence providing a series of temperature measurements in succeeding depths. It can be seen that the radiometer can be configured to operate either as a single channel or as a multi-channel one. A single branch, comprised of the same elements, can act as a single channel RF Front-end over a specific channel of bandwidth of some hundreds of MHz specified by the BPF, within the frequency range of 1-4GHz. Also, when all branches are operational, the RF Front-end accumulates measurements over several channels specified by the Band-Pass Filters (BPFs). The postulates of the suggested configuration are as follows. The SPDT microwave switch (SPDT SW) receives the (noise) signal either from the antenna or from the reference noise source. Either of the two signals is band limited via a BPF to the frequency range of interest i.e., 1-4GHz and the signal over this band is amplified by a LNA. Then, specific channels are analysed to measure the antenna's differential noise power.

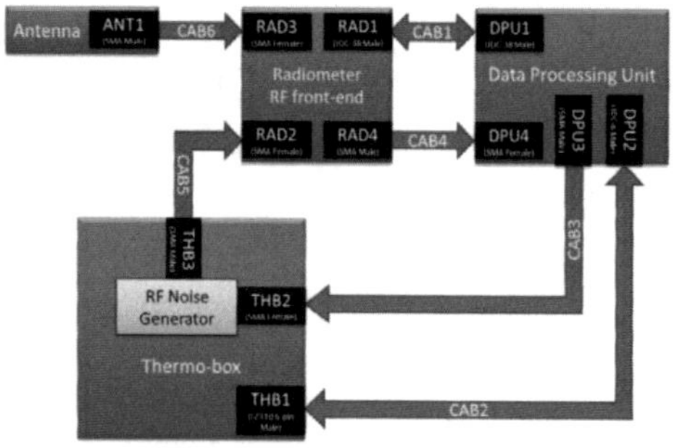

Figure 1. The Microwave Radiometer architecture

Also, a prototype "RF Reference Noise Generator" was designed and implemented based on the NC303 noise diode that delivers symmetrical white Gaussian noise and flat output power versus frequency (10MHz – 10GHz). The thermal stabilization of the noise diode NC303, is of great importance for the reliable and stable operation of this reference noise that led us to implement a thermo-box for the housing, which in combination with an embedded PID controller ensures that it is operating in a thermally stabilized environment. The proposed RF Front-end is in fact a Dicke radiometer with a channelized architecture. In this way the gain variation over the bands will be minimum and ΔT will be minimized. As for the detector, a logarithmic detector is proposed which has a far better sensitivity to -60dBm. In order to appreciate the differentiations of the proposed configuration in contrast to the typical Dicke radiometer let us define the following quantities: ΔT_{MIN} the maximum possible sensitivity in Kelvin, T_A the noise temperature of the antenna, T_E the noise temperature of the radiometer chain, B the measurement bandwidth, τ the integration time, V_{OUT} the output of the power detector, G gain of the overall radiometer chain, S the sensitivity of the square law detector, T_{Ref} is the reference noise source noise temperature and ΔT_{DICKE} is the Dicke possible sensitivity in Kelvin. Then, the typical equation that governs the Dicke radiometer is:

$$\Delta T_{DICKE} = \sqrt{\frac{2(T_A + T_E)^2 + 2(T_{Ref} + T_E)^2}{\sqrt{B \cdot \tau}} + \left(\frac{DG}{G}\right)^2 \cdot (T_E - T_{Ref})^2} \quad (1)$$

Table 1. Band Pass Filters Specifications

Filter No.	**-3dB Bandwidth (MHz)**
Band Pass Filter 1	1150-1530
Band Pass Filter 2	1600-1875
Band Pass Filter 3	1910-2275
Band Pass Filter 4	2700-2800
Band Pass Filter 5	3500-3600
Band Pass Filter 6	3100-3300
Band Pass Filter 7	2500-2575
Band Pass Filter 8	3800-3950

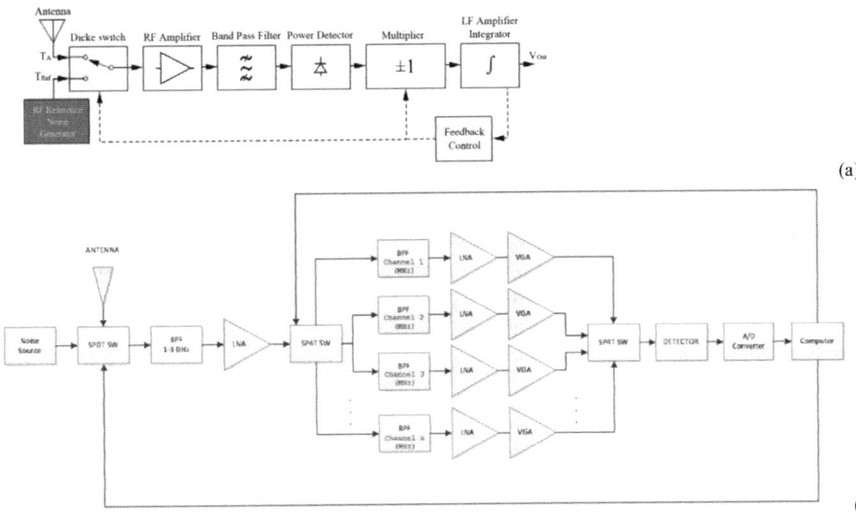

(a)

(b)

Figure 2. Block diagram of the multichannel RF Front-end: a) system level, b) analytical where channels are selected via BPFs.

It appears that if $T_E = T_{Ref}$ (which is not a realistic situation in real life radiometers), then the gain fluctuations of the radiometer chain do not play any role in the sensitivity calculations. And, in order to illustrate that the above components are adequate for the design, if we let ΔT_{MIN} = 1°C (required sensitivity), B = 200MHz, V_{OUT} = 0.5Volts (typical output of the detector), S = -50dBm (mid-range sensitivity of the detector), τ = 10msec (integration time), Ta = 200 Kelvin (antenna noise temperature), Te = 70 Kelvin (radiometer noise temperature, equivalent to approximately 4dB noise figure), T_{ref} = 100Kelvin (Reference Noise Source temperature), ΔG = 0.5dB (gain variation) and G = 40dB, then from (1) the minimum sensitivity is approximately 0.5 Kelvin which is better than the specification value of 1 degree. Of course, in the real world deployment, the expected outcome will deviate, so a margin of at least 20% in Gain "G" and Gain variation "ΔG" will be set. In our design, given the losses from the passive components, the total path achieves +18dB to +50dB gain therefore requiring about -100dBm input power from the antenna.

3. Preliminary evaluation results

Preliminary indicative results regarding the functionality as well as the evaluation of parts of the RF front-end based on off-the-shelf components using Agilent's ADS environment are given (Figure 3).

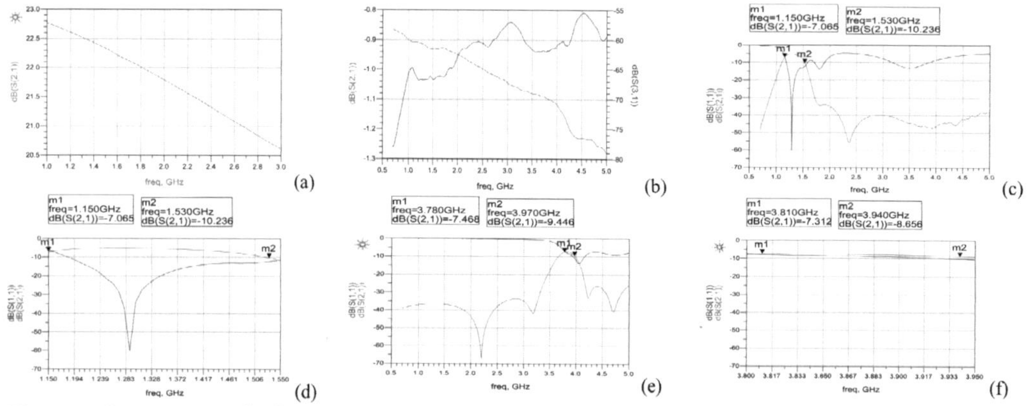

Figure 3. Measured S parameters for the, a) TSS-53LNB+ (first LNA), b) SP2T HMC349LP4CE microwave Switch, and (c)-(f) BPF channels at their extremes i.e., channel 1 (1.15-1.5GHz) and channel 2 (3.8-3.95GHz), showing full and zoomed versions to illustrate gain invariance.

As shown, the measured frequency responses (S_{21}) of the employed Switch and LNAs, depict an acceptable performance in terms of stability and gain that is expected to meet the design goals.

4. Conclusions and further work

The work in this paper has addressed some of the main issues for the design and development of the RF Front-end of a Multi-Channel Microwave Radiometer. It was identified that a successful deployment depends not only on the underlying radiometer technology, but mainly on actual implementation details and component variations and instabilities over their specifications, so as to meet the desired requirements. Ongoing work aims to implementing a prototype of the RF Front-end and integrating it to the microwave radiometer as well as performing final performance evaluation.

Acknowledgments

The research work leading to these results has been partially funded from the MMR-IBTM project (GSRT 12CHN181).

References

[1] Skou Neils 1989 *Microwave radiometer systems: design and analysis* (Norwood, MA, Artech House)

[2] Campbell A M and Land D V Dielectric properties of female human breast tissue measured in vitro at 3.2 GHz, 1992 *Physics in medicine and biology* **37**, 193.

[3] Jacobsen Svein, and Oystein Klemetsen. Improved detectability in medical microwave radio-thermometers as obtained by active antennas. 2008. *IEEE Transactions on Biomedical Engineering*, **55**, no. 12.

[4] Klemetsen A, Birkelund Y, Jacobsen S K, Maccarini P F and Stauffer P R, 2011. Design of medical radiometer front-end for improved performance. *Progress In Electromagnetics Research B*, **27**.

Journal of Physics: Conference Series **637** (2015) 012011 doi:10.1088/1742-6596/637/1/012011

Influence of Iterative Reconstruction Algorithms on PET Image Resolution

G. E. Karpetas[1], C. M. Michail[2], G. P. Fountos[2], I. G. Valais[2], D. Nikolopoulos[3], I. S. Kandarakis[2] and G. S. Panayiotakis[1]

[1] Department of Medical Physics, Faculty of Medicine, University of Patras, 265 00 Patras, Greece

[2] Radiation Physics, Materials Technology and Biomedical Imaging Laboratory, Department of Biomedical Engineering, Technological Educational Institute of Athens, Egaleo, 122 10 Athens, Greece

[3] Department of Computer Electronic Engineering, Technological Education Institute of Piraeus, Greece, Petrou Ralli & Thivon 250, 122 44, Aigaleo, Athens, Greece

e-mail: kandarakis@teiath.gr

Abstract. The aim of the present study was to assess image quality of PET scanners through a thin layer chromatography (TLC) plane source. The source was simulated using a previously validated Monte Carlo model. The model was developed by using the GATE MC package and reconstructed images obtained with the STIR software for tomographic image reconstruction. The simulated PET scanner was the GE DiscoveryST. A plane source consisted of a TLC plate, was simulated by a layer of silica gel on aluminum (Al) foil substrates, immersed in 18F-FDG bath solution (1MBq). Image quality was assessed in terms of the modulation transfer function (MTF). MTF curves were estimated from transverse reconstructed images of the plane source. Images were reconstructed by the maximum likelihood estimation (MLE)-OSMAPOSL, the ordered subsets separable paraboloidal surrogate (OSSPS), the median root prior (MRP) and OSMAPOSL with quadratic prior, algorithms. OSMAPOSL reconstruction was assessed by using fixed subsets and various iterations, as well as by using various beta (hyper) parameter values. MTF values were found to increase with increasing iterations. MTF also improves by using lower beta values. The simulated PET evaluation method, based on the TLC plane source, can be useful in the resolution assessment of PET scanners.

Keywords: PET; MTF; image quality; iterative reconstruction; Monte Carlo

1. Introduction

The most common image quality metrics in order to assess spatial resolution in Positron Emission Tomography (PET) scanners are the Point Spread Function (PSF) and the corresponding Full Width at Half Maximum (FWHM) obtained from point sources [1]. Only in a few studies spatial resolution was assessed in terms of the Modulation Transfer Function (MTF), calculated from PSF data [2]. The same method was employed in studies concerning combined micro PET/CT scanners, in which MTF was estimated in the CT detector of the combined system [3]. MTF can be also assessed from a line source through the estimation of the Line Spread Function (LSF). The use of the LSF method for determining the MTF in tomographic imagers was initially introduced by Boone [4] who applied this method for CT scanners evaluation. Fountos *et al.* [5,6] recently introduced a similar method for SPECT scanners by immersing a medical X-ray film in a solution of dithiothreitol (DTT)/Tc-99m(III)-DMSA to obtain the MTF. The purpose of the present study was to extend a previous validated Monte Carlo model [7-9] in order to obtain PET image resolution through the estimation of the MTF. Within the context of the model, MTF was estimated by the simulation of a thin layer chromatography (TLC) plane source filled with 18F-FDG, using a software model based on the GATE Monte Carlo package. The GATE package, which was developed by the Open-GATE collaboration, is an open-source extension of the Geant4 Monte Carlo toolkit [10]. GATE was used in combination with the STIR image reconstruction software [11] to obtain the plane source reconstructed images. The influence of different number of iterations and subsets, as well as the influence of various beta (hyper) parameter values, in the iterative image reconstruction was also investigated. The simulation of this plane source phantom provides an accurate model that is useful to fully characterize the performance of nuclear medicine imaging systems.

2. Materials And Methods

1.1 Geometry of the modeled PET scanner

The Discovery ST PET/CT system was modeled. The system incorporates Bismuth Germinate Oxide (BGO) [12,13] crystals with dimensions of 6.3x6.3x30 mm. The detector ring is finally comprised of 35 modules, i.e. 280 crystal blocks (for a total of 10080 BGO crystals). The dimensions of the rings are 88.6 cm diameter with a 15.7 cm axial and 70cm transaxial fields of view (FOV).

2.2. Software for simulation

The employed simulation software was the GATE toolkit. In every BGO crystal of the detector block, a Gaussian energy blur is applied with an average energy resolution of 17%, referenced at 511 keV. Afterwards, a 300 ns dead time value was applied on the single events in the BGO crystal [14] by a paralysable Deadtime module. The Quantum Detection Efficiency (QDE) [15,16] of the crystals was found 0.94 at 511 keV. The sinogram output file (.ima) was used by STIR as input file for the reconstruction of the simulated plane source image. The coincidence time window was set to 11.7 ns.

2.3. Preparation of the MTF test object

The MTF test object, which was simulated in this study, was a planar source similar to those used in previous works of Boone [4] for CT, Fountos *et al.* [5,6] for SPECT systems and Karpetas *et al.* for PET systems [9]. The plate was simulated with a layer of silica gel on Al foil substrates. The dimensions of the TLC plate were 5x10 cm^2 and it was assumed to be immersed in an 18F-FDG bath solution (1 MBq). The MTF test object was simulated between two semi-cylindrical polyethylene blocks with 20 cm diameter and 70 cm length. The importance of the hyper parameter (beta value) (which controls the weight of the penalty term and the importance of the prior) was also examined in order to investigate the corresponding influence on the MTF [17].

2.4 Modulation Transfer Function (MTF)

MTF was obtained, by using the LSF method. The thin plane source was simulated at a slight angle of 3° with respect to the horizontal or vertical axis, following the technique proposed by Fujita to avoid aliasing effects [18]. The final LSF was obtained by averaging all line LSF profiles after angle correction. The angle correction was performed following the procedure described in Fountos *et al.* [5].

3. Results and Discussion

3.2 Modulation Transfer Function (MTF)

Figure 1 shows the influence of the variation of the hyper parameter on the Maximum Likelihood Estimation (MLE)-OSMAPOSL, the Ordered Subsets Separable Paraboloidal Surrogate (OSSPS), which is a relaxed preconditioned sub-gradient descent algorithm, the Median Root Prior (MRP) and OSMAPOSL with quadratic prior, algorithms. The Median Root Prior is obtained by using Median as data processor. This prior is an extension of the idea first developed for the Median Root Prior which takes any Data Processor (i.e. a filter), and computes the prior gradient [19]. The hyper parameter showed greater influence in the MTF values when the OSMAPOSL with quadratic prior when 20 iterations was used, while the influence of beta prior was diminished with the OSSPS with 3 iterations. Figure 2 shows the resulted MTF curves when the hyper parameter is allowed to vary from a value close to zero up to 3 with various iterations and subsets, in order to investigate its impact on PET image resolution through the MTF. As it can be seen, spatial resolution is improved by using lower beta values [17]. Larger beta values result in smoother images, while lower values decrease the importance of the penalty term [17, 20]. By increasing the number of iterations, the influence of the beta parameter on image resolution tends to decrease showing smaller deviations. For example, in the 15 subsets/2 iterations image (30 MLEM equivalent iterations), deviations of the order of 60.2% are shown between the 0.01 and 3 beta value, whereas in the 300 MLEM equivalent iterations (15 subsets/ 20 iterations) image the corresponding difference decrease to 31.3%, at the spatial frequency of 0.05 cycles/mm.

Image quality in PET scanners is affected by a large number of factors including the number of iterations and subsets in ordered subset-type algorithms, filter in filtered backprojection, and post-filtering [21, 22]. In filtered backprojection reconstructed images, resolution can be assessed in a straight manner, which is not the case for OSEM images, available in most clinical systems, since iterative reconstruction algorithms are non-linear and difficult to evaluate [23]. By increasing the number of iterations, within a reasonably wide range of practically encountered clinical situations [24], resolution improves and tumors appear brighter. However, a point is reached after which resolution does not improve [21]. The hyper parameter which controls the weight of the penalty term (beta value) and the importance of the prior, has an influence also on both image resolution and noise. Spatial resolution is improved by using a lower beta value at the expense of higher image noise. In the work of Cheng *et al* [17] resolution stabilized after the value of 20 iterations. Larger values of b yield smoother images, while lower values decreases the importance of the penalty term and yield a solution closer to the MLEM algorithm [17, 20].

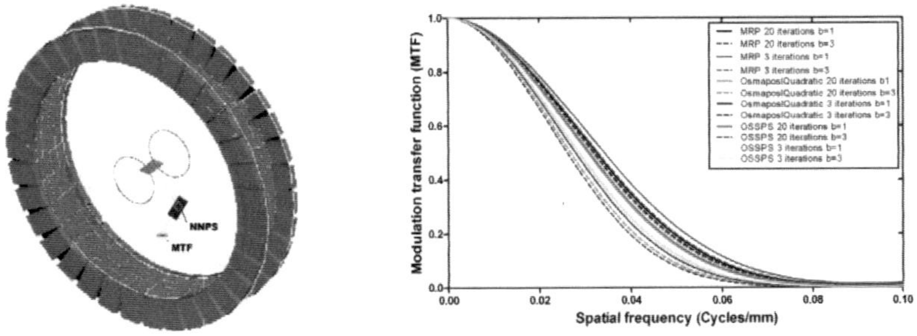

Figure 1. Simulation setup (left). Comparison between the MTFs obtained from various iterative algorithms using various values of the beta (hyper) parameter (right).

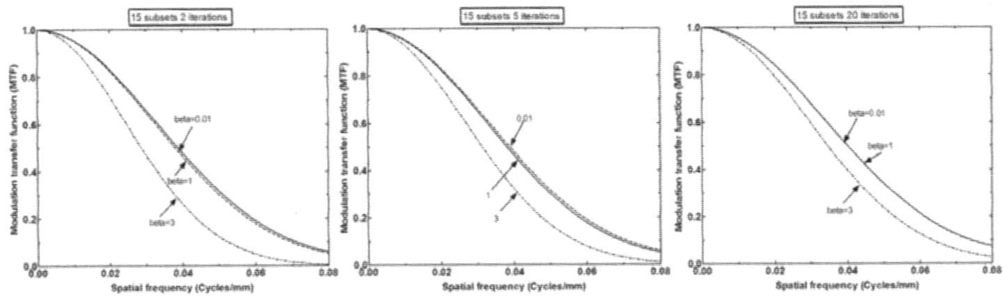

Figure 2. Comparison between the MTFs obtained from the OSMAPOSL with 30, 75 and 300 MLEM iterations (15 subsets, 2, 5 and 20 iterations) reconstructed images using various values of the hyper parameter.

4. Conclusions

Image quality characterisation of a PET scanner was achieved through the estimation of the MTF of thin 18F-FDG plane source images, simulated by a Monte Carlo model. The influence of iterative image reconstruction on the plane source simulated images was investigated by using fixed subsets and various iterations. Image quality was found to improve by increasing the number of iterations (number of image updates). Spatial resolution is also improved by using lower beta values. Larger beta values have as a result smoother images, while lower values decreases the importance of the penalty term. The method was used for the image quality assessment and optimisation, but it can be also useful in research for the further development of PET and SPECT scanners through GATE simulations.

5. References

[1] NEMA Standards Publication NU 2, 2007.
[2] Defrise M, Geissbuhler A and Townsend D W 1994 *Phys. Med. Biol.* **39** 305.
[3] Liang H, Yang Y, Yang K, Wu Y, Boone J M and Cherry S R 2007 *Phys. Med. Biol.* **52** 3881.
[4] Boone J M 2001 *Med. Phys.* **28** 356.
[5] Fountos G P, Michail C M, Zanglis A, Samartzis A, Martini N, Koukou V, Kalatzis I and Kandarakis I S 2012 *Med. Phys.* **39(3)** 1561.
[6] Fountos G, Zanglis A, Michail C, Kalatzis I, Cavouras D, Samartzis A, Kounadi E, Valsamaki P, Gerali S, Nikiforidis G and Kandarakis I, IFMBE Proceedings 25/II, pp. 802-805, 2009.
[7] Karpetas G E, Michail C M, Fountos G P, Valsamaki P N, Kandarakis I S and Panayiotakis G S 2013 *Hell J. Nucl. Med.* **16(2)** 111.
[8] Karpetas G E, Michail C M, Fountos G P, Kalyvas N I, Valais I G, Kandarakis I S, Panayiotakis G S 2014 *J. Phys.: Conf. Ser.* **490** 012139.
[9] Karpetas G, Michail C, Fountos G, Kandarakis I and Panayiotakis G 2014 *Nucl. Med. Com.* **35(8)** 967.
[10] OpenGATE Collaboration, Users Guide V6.1: http://www.opengatecollaboration.org.
[11] Thielemans K, Tsoumpas C, Mustafovic S, Beisel T, Aguiar P, Dikaios N and Jacobson M W 2012 *Phys. Med. Biol.* **57** 867.
[12] Valais I, Michail C, David S, Costaridou L, Nomicos C D, Panayiotakis G S, Kandarakis I 2008 *Physica Medica* **24** 122.
[13] Valais I, Michail C, David S, Liaparinos P, Fountos G, Paschalis T, Kandarakis I, Panayiotakis G 2010 *IEEE Trans. Nucl. Sci.* **57(1)** 3.
[14] Valais I G, Michail C M, David S L, Konstantinidis A, Cavouras D A, Kandarakis I S and Panayiotakis G S 2008 *IEEE Trans. Nucl. Sci.* **55(2)** 785.
[15] Michail C M, Fountos G P, Liaparinos P F, Kalyvas N E, Valais I, Kandarakis I S and Panayiotakis G S 2010 *Med. Phys.* **37(7)** 3694.
[16] Michail C, Kalyvas N, Valais I, David S, Seferis I, Toutountzis A, Karabotsos A, Liaparinos P, Fountos G and Kandarakis I 2013 *J. Lumin.* **144** 45.
[17] Cheng J-C, Shoghi K and Laforest R 2012 *Med. Phys.* **39(2)** 1029.
[18] Fujita H, Tsia D, Itoh T, Doi K, Morishita J, Ueda K, and Ohtsuka A 1992 *IEEE Trans. Med. Imaging* **11** 34.
[19] Thielemans K, *et al.* 2012 *STIR User's Guide Version* **2.4**, 1.
[20] Shepp L and Vardi Y 1982 *IEEE Trans. Med. Imaging* **1(2)** 113.
[21] Soret M, Bacharach S, and Buvat I. 2007 *J Nucl Med* **48** 932.
[22] Reader A J. 2008 *Physica Medica* **24** 49.
[23] Liow J, Strother S. 1993 *Phys. Med. Biol.* **38** 55.
[24] Riddell C, Carson R, Carrasquillo J, Libutti S, Danforth D, Whatley M and Bacharach S. 2001 *J. Nucl. Med.* **42** 1316.

Predicted image quality of a CMOS APS X-ray detector across a range of mammographic beam qualities

A Konstantinidis[1]
[1]Diagnostic Radiology and Radiation Protection, Christie Medical Physics and Engineering, The Christie NHS Foundation Trust, Manchester M20 4BX, UK

E-mail: Anastasios.Konstantinidis@christie.nhs.uk

Abstract. Digital X-ray detectors based on Complementary Metal-Oxide-Semiconductor (CMOS) Active Pixel Sensor (APS) technology have been introduced in the early 2000s in medical imaging applications. In a previous study the X-ray performance (i.e. presampling Modulation Transfer Function (pMTF), Normalized Noise Power Spectrum (NNPS), Signal-to-Noise Ratio (SNR) and Detective Quantum Efficiency (DQE)) of the Dexela 2923MAM CMOS APS X-ray detector was evaluated within the mammographic energy range using monochromatic synchrotron radiation (i.e. 17-35 keV). In this study image simulation was used to predict how the mammographic beam quality affects image quality. In particular, the experimentally measured monochromatic pMTF, NNPS and SNR parameters were combined with various mammographic spectral shapes (i.e. Molybdenum/Molybdenum (Mo/Mo), Rhodium/Rhodium (Rh/Rh), Tungsten/Aluminium (W/Al) and Tungsten/Rhodium (W/Rh) anode/filtration combinations at 28 kV). The image quality was measured in terms of Contrast-to-Noise Ratio (CNR) using a synthetic breast phantom (4 cm thick with 50% glandularity). The results can be used to optimize the imaging conditions in order to minimize patient's Mean Glandular Dose (MGD).

1. Introduction

Recently developed digital X-ray detectors based on Complementary Metal-Oxide-Semiconductor (CMOS) Active Pixel Sensor (APS) technology demonstrate higher performance compared to Charge-Coupled Device (CCD) and Thin Film Transistor (TFT) digital detectors. In a previous study [1] the X-ray performance of a CMOS APS X-ray detector (named Dexela 2923MAM) was experimentally evaluated within the mammographic energy range using monochromatic synchrotron radiation. In particular, the monochromatic presampling Modulation Transfer Function (pMTF), Normalized Noise Power Spectrum (NNPS), Signal-to-Noise Ratio (SNR) and Detective Quantum Efficiency (DQE) parameters were calculated for 17, 20, 23, 26, 29, 33, 33.3 and 35 keV. In the same study it was demonstrated that the monochromatic pMTF, NNPS and SNR data can be combined with different mammographic spectral shapes to predict the X-ray performance of the detector for any mammographic beam quality within the investigated energy range [1]. In the current work the monochromatic pMTF, NNPS and SNR parameters were combined with four X-ray spectral shapes. The resultant polychromatic data were inserted into an image simulation algorithm [2] to predict the respective mammographic image quality for given breast and exposure conditions.

2. Materials and Methods

Four radiation beam qualities (Molybdenum/Molybdenum (Mo/Mo), Rhodium/Rhodium (Rh/Rh), Tungsten/Rhodium (W/Rh) and Tungsten/Aluminium (W/Al) at 28 kV) were generated (using Boone's interpolating polynomial algorithm [3]) based on the IEC 62220-1-2 standard [4]. All X-ray spectra were attenuated by a synthetic average breast phantom (4 cm thick with 50% glandularity, including a cylindrical microcalcification (μC) with 600 μm diameter and 600 μm thickness). The parameters related to the attenuated X-ray spectra are shown in table 1. The used mammographic beam qualities were similar to the IEC ones (based on the Half Value Layer (HVL) values).

Table 1. Parameters related to the used radiation beam qualities for 28 kV tube voltage and 4 cm breast with 50% glandularity

Parameter	Mo/Mo	Rh/Rh	W/Rh	W/Al	Units
Anode	Mo	Rh	W	W	
Filtration	0.032 Mo	0.025 Rh	0.050 Rh	0.500 Al	mm
E_{mean}	19.4	20.9	20.9	22.2	keV
HVL	0.58	0.76	0.77	0.86	mm Al
IEC HVL	0.60	0.74	0.75	0.83	mm Al
Φ/Ka	5451	6251	6231	7087	(X-rays/mm^2/μGy)
Nom. Contrast	0.65	0.57	0.56	0.51	

Figure 1 shows the normalized spectral shapes (attenuated from the breast). Mo/Mo spectrum corresponds to lower energies, Rh/Rh and W/Rh to medium and W/Al to higher ones.

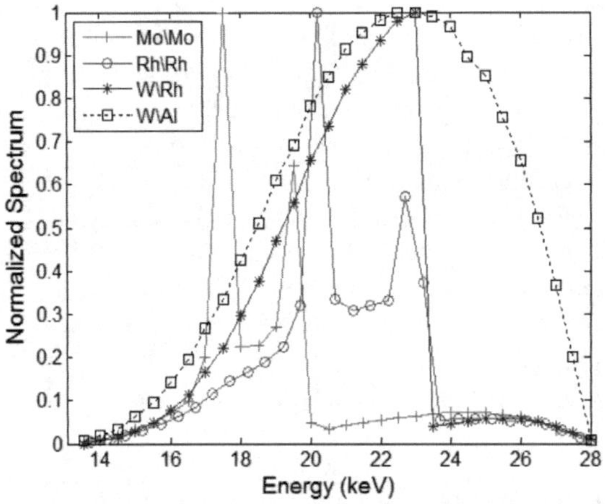

Figure 1. Normalized X-ray spectra for the used radiation beam qualities

These spectral shapes were used to calculate the polychromatic pMTF, NNPS and SNR values which were inserted in an image simulation algorithm [2] to add blurring, sampling and noise to ideal images of μC and surrounding breast tissue (with perfect uniformity). The Contrast-to-Noise Ratio (CNR) was calculated from the synthetic breast images. Figure 2 shows an ideal X-ray image, while Figure 3 shows the respective simulated one.

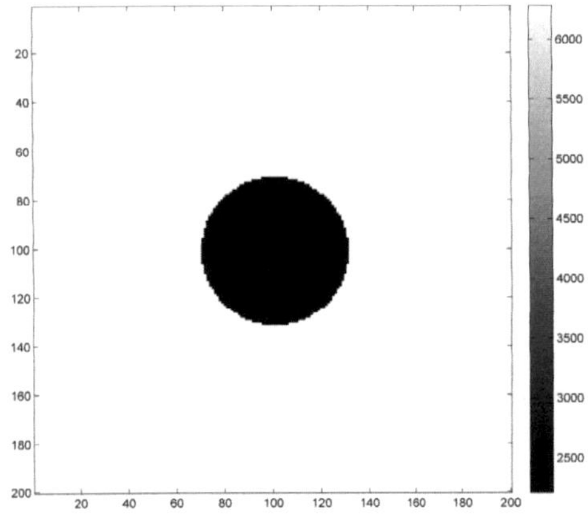

Figure 2. Ideal μC for 28 kV Mo/Mo (input)

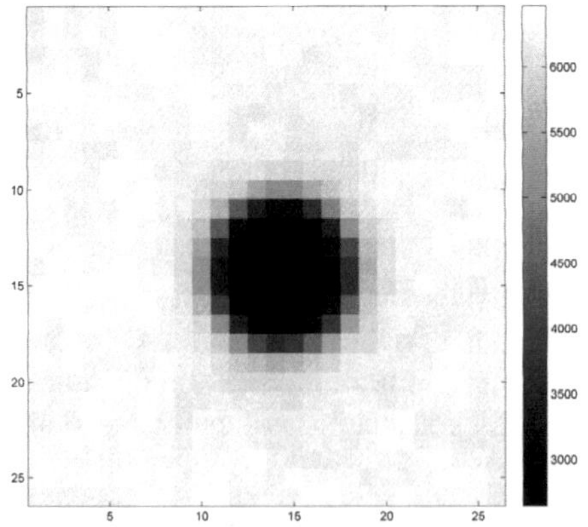

Figure 3. Simulated μC for 28 kV Mo/Mo

3. Results and Discussion

3.1. Simulation results for 3 mGy ESAK

Figure 4 shows the predicted polychromatic DQE values for 3 mGy Entrance Surface Air Kerma (ESAK), i.e. the radiation exposure at the breast entrance. Figure 5 depicts the respective CNR values. W/Rh spectrum shows the highest DQE and CNR values at the expense of high Mean Glandular Dose (MGD; see table 2), which was calculated based on [5]. This happens from the combination of the W/Rh spectral shape and the monochromatic energy deposition in the breast tissue (see [5]). On the other hand, Mo/Mo results in low Air Kerma at Detector surface (DAK), DQE and CNR values due to high attenuation from the breast. However, it demonstrates the lowest MGD value because most of the absorbed X-rays deposit their energy in the skin [5]. Finally, Rh/Rh and W/Al show similar CNR values although Rh/Rh has higher nominal contrast (see table 1) and DQE values. This probably happens because the X-ray sensitivity of the detector is higher for W/Al spectrum, resulting in improved statistics and higher SNR.

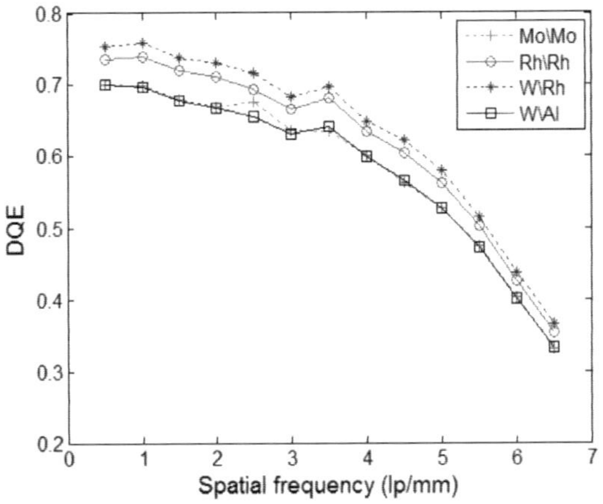

Figure 4. Predicted DQE for 3 mGy ESAK

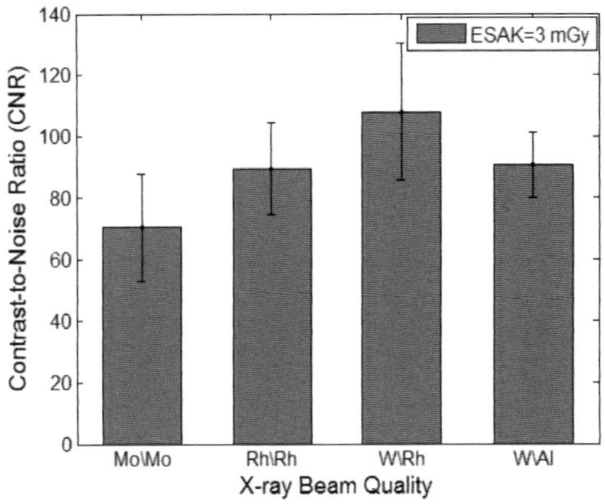

Figure 5. Predicted CNR for 3 mGy ESAK

Table 2. Calculated MGD and DAK for 3 mGy ESAK (4 cm breast with 50% glandularity)

Parameter	Mo/Mo	Rh/Rh	W/Rh	W/Al	Units
MGD	0.58	0.72	0.93	0.71	mGy
DAK	81.0	141.1	184.8	158.7	μGy

3.2. Simulation results for 100 μGy DAK

The predicted polychromatic DQE (figure 6) and CNR (figure 7) values were also calculated for an average mammographic DAK level, i.e. 100 μGy. In this case, Mo/Mo, Rh/Rh and W/Rh spectra result in high DQE values. In particular, the Mo/Mo spectrum corresponds to the highest CNR value, probably due to high nominal contrast (table 1). However, to reach 100 μGy DAK with Mo/Mo we need high ESAK (3.71 mGy) which results in high MGD (0.72 mGy) - see table 3. On the other hand, W/Rh results in high CNR and low MGD (0.5 mGy) because it requires low ESAK (1.62 mGy). It should be mentioned that the difference between the CNR values was within 13%.

Table 3. Calculated ESAK and MGD for 100 μGy DAK (4 cm breast with 50% glandularity)

Parameter	Mo/Mo	Rh/Rh	W/Rh	W/Al	Units
ESAK	3.71	2.13	1.62	1.89	mGy
MGD	0.72	0.51	0.5	0.45	mGy

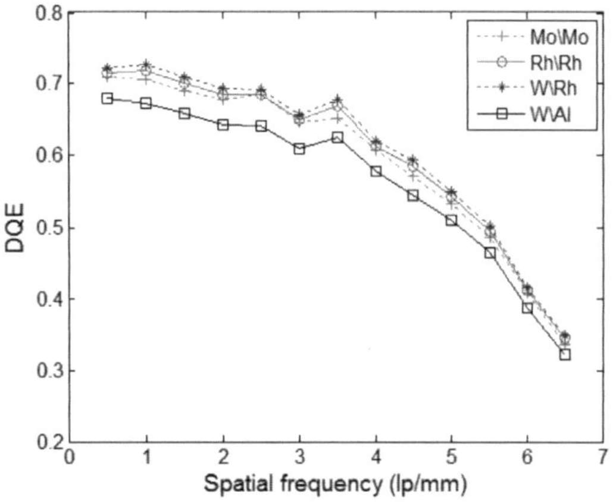

Figure 6. Predicted DQE for 100 μGy DAK

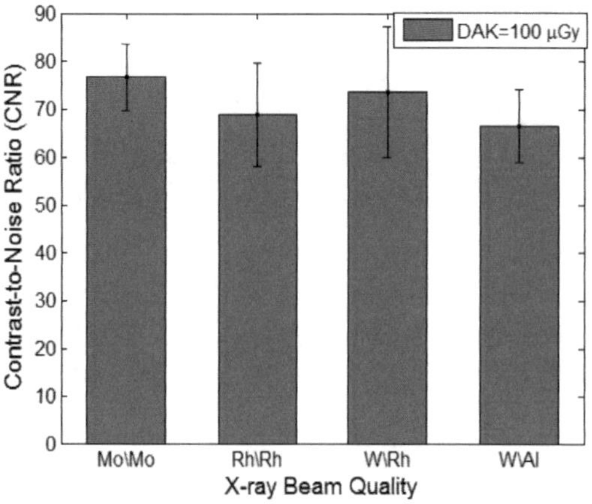

Figure 7. Predicted CNR for 100 μGy DAK

4. Conclusions and Future Work

This work demonstrates how the spectral shape affects the mammographic image quality (in terms of CNR) of a CMOS APS X-ray detector for given ESAK and DAK values. For a typical ESAK value (3 mGy) it was found that the W/Rh beam quality delivers the highest image quality at the expense of high MGD. Both Rh/Rh and W/Al spectra may deliver sufficient beam quality with lower MGD. For 100 μGy DAK the W/Rh spectrum delivers high CNR with low MGD. Further investigation needs to be made to optimize the image quality for low MGD. For example, various imaging conditions can be tested to deliver CNR values within a given range. Different X-ray spectra will be tested for various breast thickness (2-8 cm) and glandularity (0-100%) combinations. Finally, realistic software breast phantoms [6] and simulated scatter radiation effect [7] will be used to improve the simulation.

References

[1] Konstantinidis A C *et al* 2013 X-ray Performance Evaluation of the Dexela CMOS APS X-ray Detector Using Monochromatic Synchrotron Radiation in the Mammographic Energy Range *IEEE Transactions on Nuclear Science* 60 pp 3969–80

[2] Konstantinidis A C 2011 *Evaluation of digital X-ray detectors for medical imaging applications* (London: University College London)

[3] Boone J M, Fewell T R and Jennings R J 1997 Molybdenum, rhodium, and tungsten anode spectral models using interpolating polynomials with application to mammography *Medical Physics* vol 24 pp 1863–1874

[4] IEC 62220-1-2 2005 *Medical Electrical Equipment-Characteristics of Digital X-Ray Imaging Devices-Part 1-2: Determination of the Detective Quantum Efficiency-Mammography Detectors* (Geneva: International Electrotechnical Commission)

[5] Boone J M 2002 Normalized glandular dose (DgN) coefficients for arbitrary x-ray spectra in mammography: Computer-fit values of Monte Carlo derived data *Med. Phys.* 29 pp 869–874

[6] Bliznakova K, Sechopoulos I, Buliev I and Pallikarakis N 2012 BreastSimulator: A software platform for breast x-ray imaging research *Journal of Biomed. Graphics and Comp.* 2 pp 1–13

[7] Diaz O, Dance D R, Young K C, Elangovan P, Bakic P R and Wells K 2015 Estimation of scattered radiation in digital breast tomosynthesis *Phys. in Med. and Biol.* vol 59 pp 4375–4393

PVAL breast phantom for dual energy calcification detection

V. Koukou[1], N. Martini[1], K. Velissarakos[2], D. Gkremos[2], C. Fountzoula[3], A. Bakas[4], C. Michail[2], I. Kandarakis[2] and G. Fountos[2]

[1] Department of Medical Physics, Medical School, University of Patras, Rion, Patras, Greece
[2] Radiation Physics, Materials Technology and Biomedical Imaging Laboratory, Department of Biomedical Engineering, Technological Educational Institute of Athens, Egaleo, Athens, Greece
[3] Department of Medical Laboratories, Technological Educational Institute of Athens, Egaleo, Athens, Greece
[4] Medical Radiological Technology, Faculty of Health and Caring Professions, Technological Educational Institute of Athens, Egaleo, Athens, Greece

E-mail: gfoun@teiath.gr

Abstract. Microcalcifications are the main indicator for breast cancer. Dual energy imaging can enhance the detectability of calcifications by suppressing the tissue background. Two digital images are obtained using two different spectra, for the low- and high-energy respectively, and a weighted subtracted image is produced. In this study, a dual energy method for the detection of the minimum breast microcalcification thickness was developed. The used integrated prototype system consisted of a modified tungsten anode X-ray tube combined with a high resolution CMOS sensor. The breast equivalent phantom used was an elastically compressible gel of polyvinyl alcohol (PVAL). Hydroxyapatite was used to simulate microcalcifications with thicknesses ranging from 50 to 500μm. The custom made phantom was irradiated with 40kVp and 70kVp. Tungsten (W) anode spectra filtered with 100μm Cadmium and 1000μm Copper, for the low- and high-energy, respectively. Microcalcifications with thicknesses 300μm or higher can be detected with mean glandular dose (MGD) of 1.62mGy.

1. Introduction

Breast cancer is a major public health concern [1]. Microcalcifications are the principal indicator of breast cancer [2], thus the visualization and detection of microcalcifications is significant [2,3]. Dual energy (DE) imaging can enhance the detectability of calcifications by suppressing the tissue background structures [3]. With this technique, separate low- and high-energy images are acquired and a weighted subtracted image is produced.

Previous research works in dual energy imaging showed that the minimum detectable calcification size ranges from 300-355μm depending on the X-ray technique and the detection system [4,5]. Various tissue-equivalent-phantoms have been used for the experimental validation of these methods [4,5].

In this study, a dual energy imaging method is proposed for the early microcalcification detection. The prototype system used consisted of a modified tungsten (W) anode X-ray tube combined with a high resolution complementary-metal-oxide-semiconductor (CMOS) sensor. The breast equivalent phantom used was an elastically compressible gel of polyvinyl alcohol (PVAL). Contrast to noise ratio in the subtracted (dual energy) image (CNR_{DE}) was calculated for all the examined calcifications, while preserving MGD mean glandular dose (MGD) within examination levels [6].

2. Materials and Methods

Weighted log-subtraction was used to generate the DE subtraction images, $ln(DE) = ln(HE) - w\,ln(LE)$ where *HE* and *LE* are the high- and the low-energy images, and *w* is the weighting factor [7].

Experiments were carried out in a modified radiographic unit. A Del Medical Eureka X-ray tungsten (W) radiography tube was used with total inherent filtration of 3mm of aluminum (Al) [8]. The low-energy images were acquired at 40kVp with a cadmium (Cd) filter of 100μm thickness placed at the tube exit. For the high-energy images, a copper (Cu) filter of 1000μm thickness was placed in the beam at 70kVp. The detection system consisted of a terbium-activated gadolinium oxisulfide ($Gd_2O_2S:Tb$) phosphor screen (Min-R 2190 with a mass thickness of 33.91mg/cm^2) coupled to a CMOS photodiode pixel array (Remote RadEye HR) [9,10]. The CMOS photodiode array has a format of 1200x1600 pixels with a pitch of 22.5μm. The source to image receptor distance (SID) was set at 66cm. A large (1.2mm nominal) X-ray tube focal spot was selected.

The breast phantom used in the experiments of this study was an elastically compressible gel of polyvinyl alcohol (PVAL). A similar breast phantom of PVAL in ethanol and water was developed by Price *et al.* A solid yet elastically compressible gel was produced after being frozen and defrosted [11]. The linear attenuation coefficient of PVAL gels, with different concentrations ranging from 5% to 20% w/v (weight to volume), was in the range of 0.76 to 0.86cm^{-1} at 17.5keV. These values are very similar to the published breast tissue data at this energy, 0.8-0.9cm^{-1}. The breast tissue equivalent phantom used in this study was produced by mixing 50% ethanol, 50% water and 10% PVAL. The phantom thickness was 4cm. Figure 1 shows the mass attenuation coefficient of PVAL and breast tissue (ICRU-44) using published data [12]. As it can be seen from the plot, the coefficients are identical. The densities of the PVAL and breast tissue are 0.924g/cm^3 and 1.020g/cm^3 respectively, while the corresponding effective atomic numbers Z are 7.05 and 7.07 [13]. Three PMMA slabs of different thicknesses (0.2, 0.3 and 0.4cm) were used, in order to construct the phantom of microcalcifications (μC). In each PMMA slab, holes were opened and filled with a mixture of hydroxyapatite and epoxy resin (density 1g/cm^3) in order the amount of the HAp in each hole to correspond at a specific microcalcification thickness. The thicknesses of the calcifications in the μC phantoms ranged from: (i) 50 to 250μm in the 0.2cm thick μC phantom, (ii) 150 to 375μm in the 0.3cm thick μC phantom, and (iii) 100 to 500μm in the 0.4cm thick μC phantom.

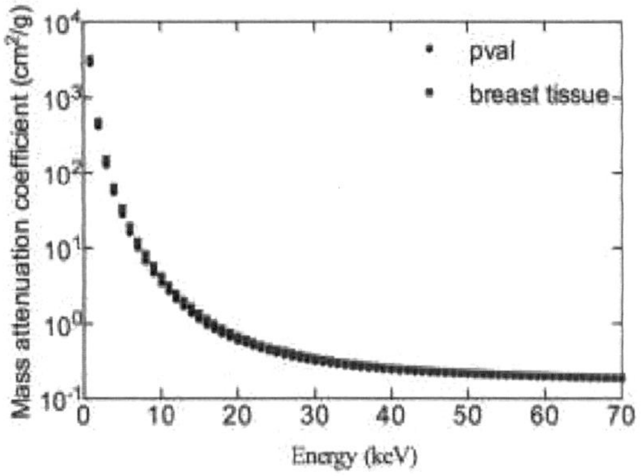

Figure 1. Mass attenuation coefficients of PVAL and breast tissue (ICRU-44).

CNR was used as a measure of image quality and is a function of both the low- and high-energy images. The CNR_{DE} was calculated according to the following equation [14]:

$$CNR_{DE} = \frac{\left|\overline{S_{\mu C}} - \overline{S_B}\right|}{\sigma_B} \quad (1)$$

where μC and B denote the mean signal of the microcalcification and background regions, and σ represents the standard deviation of pixel values in the background [14]. The CNR_{DE} threshold value for the detection of a calcification was equal to 3 [3].

The MGD was calculated using the practical relation of ACR [15] and previously published data [16].

3. Results and Discussion

Figure 2 shows the dual energy images of homogenous PVAL phantom combined with the examined three-calcification phantoms. The entrance dose was 1.92mGy, corresponding to MGD value of 1.62mGy.

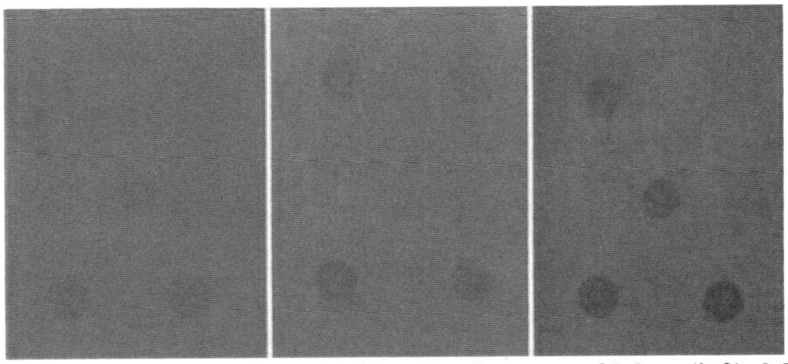

Figure 2. Dual energy images of the three-calcification phantoms of 0.2cm (left), 0.3cm (center) and 0.4cm (right) thick.

Table 1 shows the measured CNR_{DE} values of the three-calcification phantoms, with thicknesses ranging from 150 to 500μm for MGD values of 1.62 and 0.80mGy. Calcification thicknesses of 50 and 100μm could not be depicted in the DE image due to: (i) counting statistics leading to increase of image noise and (ii) low object contrast which was expected from the simulation study [17]. For this reason, CNR_{DE} values were not included in Table 1. The minimum detectable calcification thickness was 300μm, as yielded a CNR_{DE} value of 3.25 which is above the threshold of 3. In a previous dual energy study, the minimum detectable calcification size was reduced to 250μm after applying noise reduction techniques in the DE images [5]. However, in this study, noise reduction methods were not included.

Table 1. Measured CNR_{DE} values of the three-calcification phantoms.

Calcification thickness (μm)	CNR_{DE} (MGD=1.62mGy)	CNR_{DE} (MGD=0.80mGy)
150	1.71	0.79
200	2.11	1.39
225	2.53	1.41

250	2.58	1.97
300	3.25	2.31
375	3.38	2.56
400	3.71	2.65
500	4.63	3.01

4. Conclusions

In this study, a homogenous breast-equivalent-phantom was developed using polyvinyl alcohol, water and ethanol. A mixture of hydroxyapatite and epoxy resin was used for microcalcifications with various thicknesses. Contrast to noise ratio was calculated from the DE subtracted images. The minimum detectable calcification thickness was 300μm for mean glandular dose of 1.62mGy.

Acknowledgement

This research has been co-funded by the European Union (European Social Fund) and Greek national resources under the framework of the “Archimedes III: Funding of Research Groups in TEI of Athens” project of the “Education & Lifelong Learning” Operational Programme.

References

[1] World Health Organization 2009 Available via: http://www.who.int/gender/women_health_report/full_report_20091104_en.pdf
[2] Mou X and Chen X 2007 *Proc of MICCAI (Australia)* vol 4792 p 596
[3] Lemacks M, Kappadath S, Shaw C, Liu X, Whitman G 2002 *Med. Phys.* **29** 1739
[4] Kappadath S and Shaw C 2005 *Med. Phys.* **32** 3395
[5] Kappadath S and Shaw C 2008 *Phys. Med. Biol.* **53** 5421
[6] EUREF 2006 European guidelines for quality assurance in breast cancer screening and diagnosis 4th ed European Commission
[7] Chung H, Ikejimba L, Kiarashi N, Samei E, Hoernig M and Lo J *Proc. SPIE 8668, Medical Imaging 2013: Physics of Medical Imaging* **86684L**
[8] Del Medical Systems Group, Roselle, IL [Online]. Available from: http://www.delmedical.com.
[9] Michail C, Spyropoulou V, Fountos G, Kalyvas N, Valais I, Kandarakis I and Panayiotakis G 2011 *IEEE Trans. Nucl. Sci.* **58** 314
[10] Seferis I, Michail C, Valais I, Fountos G, Kalyvas N, Stromatia F, Oikonomou G, Kandarakis I, Panayiotakis G. 2013 *Nucl. Instrum. Meth. Phys. Res. A.***729** 307
[11] Price BD, Gibson AP, Tan LT and Royle GJ 2010 *Phys. Med. Biol.* **55** 1177
[12] Boone JM and Chavez AE 1996 *Med. Phys.* **23**, 1997
[13] Billas I, 2011, Angiogenesis measurements in mammography using time-resolved dual energy analysis, MSc thesis, University of Patras. Available online: http://nemertes.lis.upatras.gr/jspui/handle/10889/4869
[14] Ducote JL, Xu T and Molloi S 2007 *Phys. Med. Biol.* **52** 183
[15] ACR 1999 Mammography, Quality Control Manual (Reston, VA: Americal College of Radioliogy)
[16] Boone J 1999 *Radiol.* **213** 23
[17] Koukou V, Martini N, Michail C, Sotiropoulou P, Fountzoula C, Kalyvas N, Kandarakis I, Nikiforidis G, Fountos G. 2015 *Comput. Math. Methods Med.* [in print]

Organ dose and risk assessment in paediatric radiography using the PCXMC 2.0

A Ladia, G Messaris, H Delis, G Panayiotakis

Department of Medical Physics, School of Medicine, University of Patras, Patras,26500, Greece

E-mail: panayiot@upatras.gr

Abstract. Abdominal and chest radiographs are the most common examinations in paediatric radiology. X-ray examination of children attracts particular interest, mainly due to the increased risk for the expression of delayed radiogenic cancers as they have many years of expected life remaining. This study aims to calculate the organ dose and estimate the radiation Risk of Exposure Induced cancer Death (REID) to paediatric patients, using the PCXMC 2.0 Monte Carlo code.Patient data and exposure parameters were recorded during examinations of 240 patients, separated in four age groups undergoing chest or abdomen examinations.The organs received the highest dose in all patient groups were liver, lungs, stomach, thyroid, pancreas, breast, spleen in chest radiographs and liver, lungs, colon, stomach and ovaries, uterus (for girls) and prostate (for boys) in abdomen radiographs. The effective dosefor the chest was 0.49 $x10^{-2}$- $1.07x10^{-2}$ mSv, while for the abdomen $1.85x10^{-2}$- $3.02x10^{-2}$ mSv. The mean REID value was1.254 $x10^{-5}$ for the abdomen and$0.645x10^{-5}$for the chest.

1. Introduction

Abdominal and chest radiographs are the most common examinations in paediatric radiology. X-ray examination of children attracts particular interest, mainly due to the increased risk for the expression of delayed radiogenic cancers, as they have many years of expected life remaining. Additionally, their smaller sizes bring all organs within or closer to the useful x-ray field, resulting to a higher overall exposure per radiograph [1].

Entrance surface dose (ESD) is one of the dosimetric quantities used in radiography and is useful to determine the level of patient doses, though it is not directly related to risk [2]. The evaluation of risk to paediatric patients requires the dose for various tissues and organs. One common method for evaluating organ doses is based on measurements using thermoluminescence dosemeters (TLD). TLD dosimetry, however, is laborious and time consuming [3-4]. Another method for estimating organ dose is Monte Carlo (MC) simulation. Monte Carlo simulation is a method applied extensively. The PCXMC code was developed at the Medical Radiation Laboratory of the Finnish Radiation and Nuclear Safety Authority for calculating patient doses in diagnostic radiology utilized relative simple measurements [5].

This study aims to calculate the mean value of the organ and effective doses, using the Monte Carlo PCXMC 2.0 code and estimate the radiation risk for paediatric patients in routine radiographic examinations. The calculation of the dose to each organ separately, in combination with the effective dose, provides an estimate of radiation risk.

2.Materials and methods

2.1.Measurements

Exposure parameters (tube load, tube voltage, tube current, Field Of View (FOV), Focus Skin Distance (FSD) and patient data (sex, age, weight, and height) were recorded for each examination at the Karamandaneio Children Hospital of Patras, using the GE Model MS 18S radiology unit with tube filtration 3.5 mm Al. In this study, 120 paediatric patients underwent chest examination and 120 abdominal examinations. The chest radiographs were posterior-anterior and the abdomen radiographs anterior-posterior projection. No special preparation took place before examinations.Children were selected so that their anatomical characteristics (weight and height) to be representative of age groups (1, 5, 10, 15 years old) according to the requirements of the code used. The nominal anatomical data were based on the mathematical hermaphrodic phantom models of Christy and Eckerman [6-9].

2.2. Monte Carlo simulation
The PCXMC 2.0 code was used to calculate the dose of each organ separately and the effective dose of chest and abdomen x-ray examinations.All doses were calculated based on the exposure and patient data recorded. Photons were emitted isotropically from a point source into the solid angle specified by the FSD and FOV [6-9].The simulations were made using Intel ® Core™ 2 Duo processor of 2.66 GHz CPU powered by Asus and 4 GB installed memory (RAM). The number of x-ray photons involved in each simulation was $2\cdot 10^6$ in order to assure a low relative statistical uncertainty of 0.1 %. The time required for each simulation was 20 to 30 minutes.

2.3. Effective doses and risk assessment.
The effective dose, for each radiograph examination, was computed according to the equation[10]:

$$E=\sum_T W_T \left[\frac{H_T^M+H_T^F}{2}\right] \quad (1)$$

where$W_T$is the tissue weighting factor and H_T^M,H_T^F the equivalent doses for tissue T of male and female.

For the assessment of cancer risk resulting from an exposure to ionizing radiation [Risk of Exposure Induced cancer Death (REID)], the BEIRVII mathematical model was used, for all age groups. The REID values were calculated using the formula[7-8]:

$$REID_C\,(\mathrm{e,\,D})=\int_T^{\infty}[\,\mu_c(t|e,D)\text{-}\mu_c(\mathrm{t})]\;\mathrm{S}\;(t|e,D)dt \quad (2)$$

where$\mu_c(t|e,D)$ is the mortality rate at age t due to death cause c, given that the subject was alive at the age of exposure e and the corresponding dose at the age was D, μ_c(t) is the background mortality rate related to the death because c, S$(t|e,D)$is the conditional probability that the subject is alive at age t, given a dose D at the age of exposure e.

3. Results and discussion

The patient data and the exposure parameters for chest and abdomen radiographs are presented in Table 1 and 2, respectively.

Table 1. Patients data and exposure parameters for chest radiographs.

Age (y)	weight (kg)		height (cm)		FSD (cm)	Tubevoltage (kVp)		Tubeload (mAs)	
	Range	Mean	Range	Mean	Mean	Range	Mean	Range	Mean
1	6.80-13	9.60	63-100	76.70	122	75-80	77.70	1.12-6.24	2.60
5	12.50-22	16.70	89-120	100.70	151	72-82	79.10	1.89-6.89	3.89
10	31-38	30.80	95-150	129.70	147	77-98	80.30	2.02-9.41	5.40
15	42-76	52.20	140-178	156.40	158	78-95	82.30	2.06-9.91	5.66

Table 2. Patients data and exposure parameters for abdomen radiographs.

Age (y)	weight (kg)		height (cm)		FSD (cm)	Tubevoltage (kVp)		Tubeload (mAs)	
	Range	Mean	Range	Mean	Mean	Range	Mean	Range	Mean
1	8-13.50	10.26	62-90	79.60	118	70-80	75.50	3.21-12.50	6.21
5	14-26	19.40	92-120	111.00	151	70-80	74.40	3.89-17.20	7.86
10	27-39	32.23	120-145	135.60	150	70-80	76.60	3.95-18.25	10.04
15	40-61	49.20	145-165	151.10	160	70-80	75.50	8.89-22.60	15.43

Table 3. Organ doses in chest examinations.

Organ	Dose (mGy) x 10^{-2}			
	1 y	5 y	10 y	15 y
Liver	4.10	3.23	3.28	2.66
Lungs	4.71	4.66	6.07	5.84
Stomach	4.23	7.82	2.10	1.61
Thyroid	4.57	1.35	1.00	0.61
Pancreas	4.04	3.12	3.12	2.50
Breast	1.96	1.77	2.42	1.00
Spleen	3.15	2.98	2.95	2.35

Table 4. Organ dosesin abdomenexaminations.

sex	Organ	Dose (mGy) x 10^{-2}			
		1 y	5 y	10 y	15 y
	Liver	1.61	1.40	0.96	1.91
	Lungs	0.98	0.59	0.59	0.52
	Stomach	3.91	3.89	3.31	3.31
	Colon	3.33	3.20	3.03	3.74
girls	Ovaries	3.07	3.03	3.50	3.03
	Uterus	3.35	3.36	3.66	3.29
boys	Prostate	3.09	3.25	3.34	3.28

Table 5. Mean Effective dose for chest and abdomen examinations.

Age (y)	Mean Effective Dose (mSv) x 10^{-2}	
	chest	abdomen
1	0.94	3.02
5	1.07	1.26
10	0.49	2.11
15	0.98	1.85

Table 6. Mean REID values for all patient groups for both examinations.

	REID x10^{-5}
Chest	0.645
Abdomen	1.254

Tables 3-6 present the organ dose, effective dose and the REID values, according to ICRP 103, for each examination and age group. The dose contribution analysis for all age groups indicates that some organs received greater dose amounts, such as liver, lungs, stomach, thyroid, pancreas, breast, spleen in chest radiographs and liver, lungs, colon, stomach and ovaries, uterus (for girls), prostate(for boys) in abdomen radiographs. The dose of the other organs was negligible.

4. Conclusions

The organs received the highest doses in all patient groups studied were liver, lungs, stomach, thyroid, breast, pancreas, spleen in chest radiography and liver lungs, colon, stomach and ovaries, uterus (for girls) and prostate (for boys) in abdomen radiography.The value of mean effective dose was up to four times higher for abdomen compared to chest. The value of REID for girls was slightly greater than boys for each examination. The mean REID value was almost double(1.254x10^{-5})for abdomencompared to chest(0.645x10^{-5}). The risk was extremely low for both chest and abdomen examinations for all patient groups studied.

References

[1] Committee on Enviromental Health 1998*Pediatrics* **101** 717-719.
[2] Ribeiro L.A., Yoshimura E.M. 2008*Rad. Meas.***43**972–976.
[3] Yakoumakis E. et al 2007 *Rad. Prot. Dosim.* **124** 348-52.
[4] Kiljunen T et al 2009 *Acta Rad.* **50** 114-124.
[5] Servomaa A. Tapiovaara M. 1998 *Rad. Prot. Dosim.* **80** 213-219.
[6] Eckerman, K. F., Cristy, M. and Ryman, J. C. 1996 *The ORNL mathematical phantomSeries*. Available on http://ats.ronl.gov/documents/mird2.pdf.
[7] International Commission on Radiological Protection (ICRP). Basic for dosimetric quantities in radiological protection. Annex B of main Recommendations. Publication 103 2006 Elsevier.
[8] BEIR - Committee to Access Health Risk from exposure to low levels of ionizing radiation. BEIR VII. 2006 National Academy of Science.
[9] Tapiovaara M, SiiskonenT. PCXMC A montecarlo program for calculating patient doses in medical x-ray examination (2nd Ed.) 2008 STUK-A231 Helsinki, Finnish Centre for radiation and nuclear safety.
[10] Yakoumakis E et al 2015 *Rad. Prot. Dosim.***163** 202-209.

Hybrid Pixel Detectors for gamma/X-ray imaging

D. Hatzistratis[1,2], G. Theodoratos[1,3], V. Zografos[1,4], I Kazas[5], D. Loukas[5], and C. P. Lambropoulos[1]
[1]Aircarft Engineering Department, Technological Educational Institute of Sterea Ellada, Greece
[2]Ajat Oy Ltd., Finland
[3]Broadcom Corporation, USA
[4]Department of Electrical and Computer Engineering, National Technical University of Athens, Greece
[5] Institute of Nuclear Physics, National Center for Scientific Research Demokritos, Greece

E-mail: lambrop@mail.teiste.gr

Abstract. Hybrid pixel detectors are made by direct converting high-Z semi-insulating single crystalline material coupled to complementary-metal-oxide semiconductor (CMOS) readout electronics. They are attractive because direct conversion exterminates all the problems of spatial localization related to light diffusion, energy resolution, is far superior from the combination of scintillation crystals and photomultipliers and lithography can be used to pattern electrodes with very fine pitch. We are developing 2-D pixel CMOS ASICs, connect them to pixilated CdTe crystals with the flip chip and bump bonding method and characterize the hybrids. We have designed a series of circuits, whose latest member consists of a 50x25 pixel array with 400um pitch and an embedded controller. In every pixel a full spectroscopic channel with time tagging information has been implemented. The detectors are targeting Compton scatter imaging and they can be used for coded aperture imaging too. Hybridization using CMOS can overcome the limit put on pixel circuit complexity by the use of thin film transistors (TFT) in large flat panels. Hybrid active pixel sensors are used in dental imaging and other applications (e.g. industrial CT etc.). Thus X-ray imaging can benefit from the work done on dynamic range enhancement methods developed initially for visible and infrared CMOS pixel sensors. A 2-D CMOS ASIC with 100um pixel pitch to demonstrate the feasibility of such methods in the context of X-ray imaging has been designed.

1. Introduction

Compound semiconductor sensors, mainly CdTe/CdZnTe, find steadily increasing application in room – temperature, large volume X/Gamma-ray imaging coupled with spectrometry in the security [1], medical[2], industrial[3] and space sectors[4] with driving force the medical sector. These sensors are converting the impinging gamma photon to electron-hole pairs with satisfactory efficiency due to their atomic number (48,52). Direct conversion is attractive because it exterminates all the problems of spatial localization related to light diffusion. The energy resolution of compound semiconductor sensors (less than 1% @511KeV has been reported for pixelated CdTe detectors for PET applications) is far superior from the combination of scintillation crystals and photomultipliers (10-15%

@511KeV). Furthermore lithography can be applied to pattern electrodes with very fine pitch and thus result to position sensitive detectors with sub-millimeter spatial resolution at a large scale.
Two developments with the use of hybrid pixel technology will be presented. The one targets the localization of radioactive sources and the other targets wide dynamic range X-ray imaging.

2. A hybrid pixel detector for the localization of radioactive sources

The first hybrid pixel detector discussed aims to provide the information necessary for the identification and localization of radioactive sources. The identification poses the requirement to provide spectroscopic information. The localization can be performed using the coded aperture [5], or the Compton imaging technique [6].
Both methods require spatial resolution, which means that the address of the pixels having signal should be recorded. Additionally Compton imaging requires time tagging capability, because the determination of the correct time sequence of interactions is necessary for the efficient reconstruction of the incident direction of the photons.
The hybrid shown in figure 1(b) has detecting layer made of 2mm thick CdTe with "ohmic type" electrodes. The pixel pitch is 400um. Part of the pixel array of the CMOS ASIC designed for this detector is shown in figure 1(a), together with the bumps developed on the CMOS surface to connect the electronic circuits' inputs to the CdTe crystal pixel electrodes. The CdTe/CdZnTe crystals with "ohmic type" electrodes function like solid state ionization chambers.

Figure 1. (a) Bi/Sn bumps developed on the in-pixel pads, (b) The hybrid detector module.

The same hybrid has been constructed by using 0.75mm thick CdTe crystal but with electrodes which create Schottky type diodes on the surface between the pixel electrode and the CdTe bulk.
Indicative spectra recorded from different pixels of the hybrid are shown in figure 2. The horizontal axis is the voltage output in Volts of the in-pixel peak detector, which is proportional to the charge collected in the pixel. One can notice that in every pixel the counts start from a different value of the horizontal axis. This is due to the fact that the hit flag comparator threshold is not set globally, but it is set by design in each pixel separately to be 20mV above the baseline of the in-pixel shaper. Also the gain of each pixel is different as it can be seen by the varying distance between the ^{241}Am and the ^{57}Co peaks.
By fitting two Gaussian curves with appropriate cuts to the spectra of the (col 4, row 33) pixel one can approximate the gain to be 1.038KeV/mV and the FWHM to be 6.15KeV for the ^{241}Am peak and 8.24KeV for the ^{57}Co peak. The same procedure performed on another pixel results to 3.38KeV FWHM for the ^{241}Am peak and to 7KeV for the ^{57}Co peak. Noise measurements show an equivalent

noise charge from 400 to 500 electrons for the same gain settings for which the spectra from radioactive sources were recorded. A vast amount of characterization work is underway.

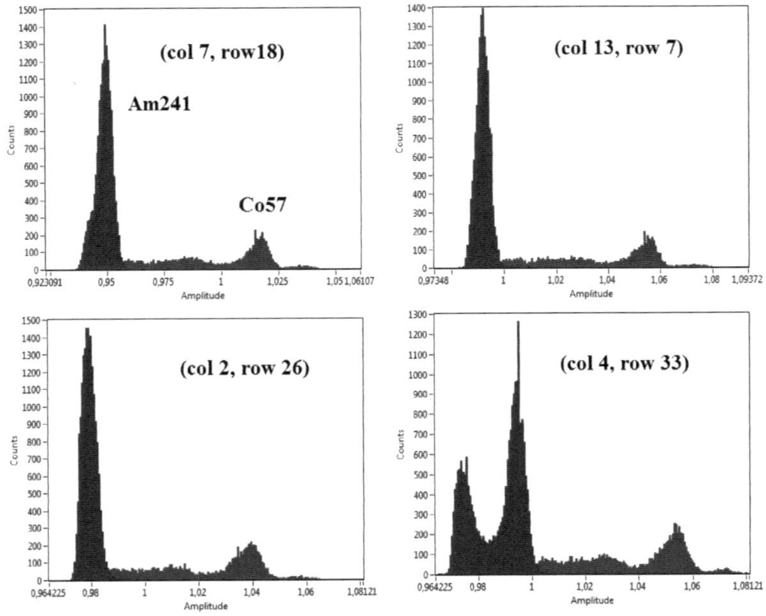

Figure 2. Pixel spectra recorded with the hybrid detector: Al/CdTe/Pt + P4DI ASIC

3. A high dynamic range hybrid pixel sensor

Photon counting in contrast to charge integration requires complex pixel electronics and inevitably more power consumption. In large area X-ray imaging the integrated photocurrent induced during exposure is measured. The TFT transistor technology used in flat panels does not permit the encapsulation of in-pixel intelligence which can be implemented easily in CMOS technology. For this reason first CMOS passive and subsequently active pixel sensors have found fertile ground for application in X-ray imaging. The photocurrent is integrated on a capacitance, whose voltage saturates. This problem has been first confronted in the development of CMOS pixel sensors for visible and IR light and a rich set of solutions has been devised after intense research work [7]. Dental intraoral and extra-oral radiography, dental computed tomography, 3-D cephalometry and industrial computed tomography are possible candidate applications for the adoption of these techniques. The method we implement is called folded multiple captures [8] and is explained below with the aid of an example. Self-reset of the integrating capacitance is used in order to avoid saturation during the frame integration. A threshold of 1.8V sets the limit for the voltage, which can be built on the integrating capacitance. In the first row of figure 3(a) the photo current is integrated on a capacitance with initial voltage value 0.5V for a frame time of 1ms producing a voltage vphoto. At the same time a second ramp voltage (vtime) increases proportional to time. The two voltages are sampled at n (in the example n=4) predefined times. Consequently during one frame n pairs of (vphoto, vtime) are read out from one pixel. These n data pairs are used for a linear fit whose slope is proportional to the photocurrent. In the second row of figure 3(a) a higher photocurrent is integrated. Here vphoto reaches the 1.8 V threshold after 0.4ms, is reset to the initial value of 0.5V and integrated again to reach the threshold for a second time after 0.8ms. The voltage ramp vtime is reset simultaneously with the integrating capacitance. Thus when (vphoto,vtime) are captured in the predefined times, vtime is proportional to the time since last reset. Also the vphoto is the voltage built on the integrating capacitance since last reset. Thus the data pairs can be fitted to give a higher slope which is the estimated photocurrent.

The block diagram of the pixel readout circuit is depicted in figure 3(b). Two channels exist in each pixel. The one records the photo-current and the other records time. Time from last reset is recorded by connecting a global ramp signal to the one node of a capacitor, whose other node is driven to a source follower input buffer. Source followers are the input and output buffers of the sample and hold circuits, which capture vphoto and vtime. Self-reset is implemented by a comparator with a global threshold. Thus, two analog voltage levels are captured and readout in every pixel.

(a)

(b)

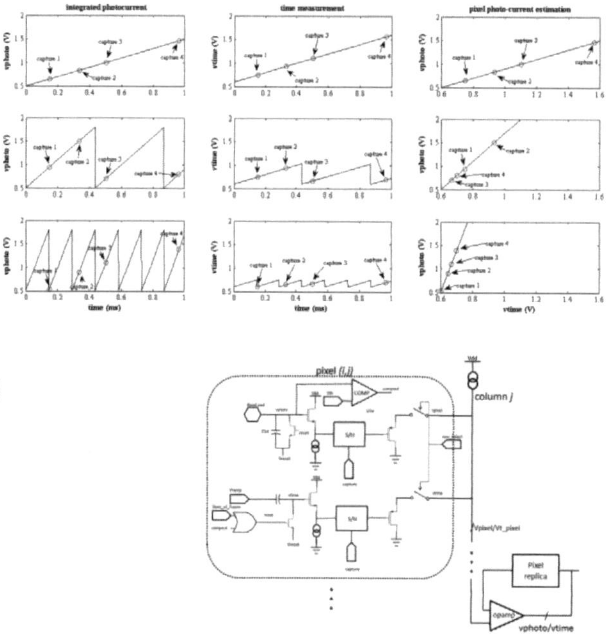

Figure 3. (a)Folded multiple captures principle of operation.(b) pixel architecture

Acknowledgments
The research has been funded by the FP7-SEC-218000 COCAE, the ESA C21213/NL/PA Contract and by the European Union (European Social Fund – ESF) and Greek national funds through the Operational Program "Education and Lifelong Learning" of the National Strategic Reference Framework (NSRF) - Research Funding Program: ARCHIMEDES III. Investing in knowledge society through the European Social Fund.

4. References

[1] M.J. Myjak, C.E. Seifert 2008 *IEEE Trans. Nucl. Sci.* **55** 769
[2] C. Tanaka, H. Fujii, A. Shiotani, Y. Kitagawa, K. Nakamura, A. Kubo, 2005 *Clinical Nuclear Medicine* **30** 440
[3] B Blakeley 2006 *Insight-Non-Destructive Testing and Condition Monitoring* **48** 109
[4] O. Limousin et al., 2011 *Nucl. Instr. and Meth. in Phys. Res.* A **647** 46
[5] E. Caroli, J. B Stephen, G. di Cocco, L. Natalucci, and A. Spizzichino 1987 *Space Science Reviews* **45** 349
[6] T. Kamea, R. Enomoto, and N. Hanada 1987 *Nucl. Instr. and Meth. in Phys. Res.* A **260** 254
[7] A. Spivak, A. Belenky, A. Fish and O. Yadid-Pecht, 2009 *IEEE Trans. Electron Devices* **56**, 2446
[8] S. Kavusi and A. El Gamal "Folded Multiple-Capture: An Architecture for High Dynamic Range Disturbance-Tolerant Focal Plane Array", in *Proc. of SPIE* vol. 5406, p 351.

3-D localization of gamma ray sources with coded apertures for medical applications

I Kaissas[1], C Papadimitropoulos[2], K Karafasoulis[3], C Potiriadis[1] and C P Lambropoulos[2]
[1]Greek Atomic Energy Commission, Greece
[2]Technological Educational Institute of Sterea Ellada, Greece
[3]Hellenic Army Academy, Greece

E-mail:lambrop@mail.teiste.gr

Abstract. Several small gamma cameras for radioguided surgery using CdTe or CdZnTe have parallel or pinhole collimators. Coded aperture imaging is a well-known method for gamma ray source directional identification, applied in astrophysics mainly. The increase in efficiency due to the substitution of the collimators by the coded masks renders the method attractive for gamma probes used in radioguided surgery. We have constructed and operationally verified a setup consisting of two CdTe gamma cameras with Modified Uniform Redundant Array (MURA) coded aperture masks of rank 7 and 19 and a video camera. The 3-D position of point-like radioactive sources is estimated via triangulation using decoded images acquired by the gamma cameras. We have also developed code for both fast and detailed simulations and we have verified the agreement between experimental results and simulations. In this paper we present a simulation study for the spatial localization of two point sources using coded aperture masks with rank 7 and 19.

Introduction

In cancer surgery, intraoperative localization of spots of radio-traced tissues is implemented by gamma counter probes or more sophisticated handheld gamma cameras [1]. For instance, sentinel nodes with diameter from 0.8 to 1.5 cm are traced with ^{99m}Tc injection in order to identify and remove the cancerous ones with intraoperative use of handheld detectors [2]. Due to the physiology of lymphs and pharmacokinetics, ^{99m}Tc occupies only a small portion of the entire node. Also due to the anatomy of lymphatic system, these nodes are spaced apart more than a couple of centimeters. Considering ^{99m}Tc concentrations as point-like sources, in this study we investigate the use of two CdTe pixelated detectors with coded apertures and the combination of their images with the triangulation method in order to locate point-like radioactive sources in 3-D space. Application in radioscintigraphy of one gamma camera with a coded mask has been investigated recently [3].

Materials, definitions and methods

Simulations and measurements are employed to evaluate the imaging capabilities of an integrated system that consists of two energy dispersive and position sensitive detectors and a coded aperture (or mask) mounted on top of each detector. The mask-to-detector distance b and the spacing between the two detectors d are adjustable with b_{min}=18mm and d_{min}=17cm due to the detector housing. These two parameters (b_{min}, d_{min}) specify a minimum distance for normal operation (z_{min}=7.82cm) where the intraoperative procedure can record a clear signature of the source on both detectors.

Each position sensitive detector is an array of 4×2 hybrids made by pixelated CdTe chips, 1mm thick, bump-bonded to pixel readout electronics in the form of Application Specific Integrated Circuit (ASIC) which record the charge delivered to each pixel. The active area of the detector is 4.4×4.4 cm^2 and has 16384 pixels with 350um pitch. The system can record signals from 15keV up to 200keV with 1KeV step and with 3-4 keV FWHM energy resolution as reported in [4] and verified by measurements with ^{241}Am, ^{57}Co and ^{99m}Tc. The absorption efficiency of the 1mm thick CdTe at 140KeV is 30%. The results reported below are from data both experimental and simulated taken with a 10mCi ^{241}Am source 1mm in diameter.

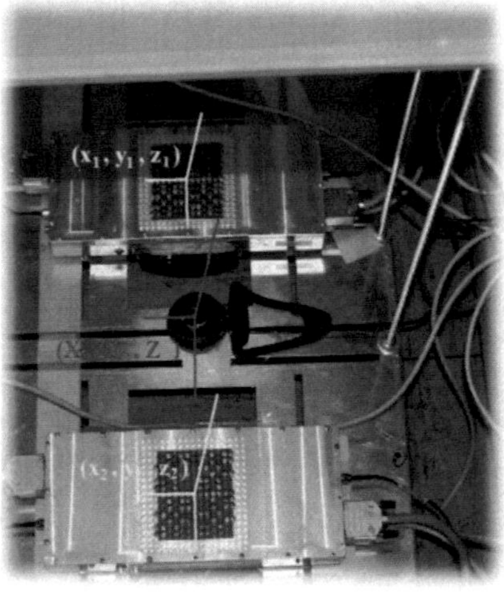

Figure 1. A view of the system. The two gamma cameras with coded apertures mounted on an aluminum base, can be seen together with a web-camera. The image also depicts the co-ordinate systems (x_i, y_i, z_i), i=1,2 of the two cameras used to extract the displacement of their SPSF, as well as the system reference frame (X,Y,Z).

For the experimental setup (figure 1) we have constructed two pairs of MURA coded apertures with Ranks 7 and 19, respectively. In Table 1 their main characteristics are outlined. Both masks have been constructed on a PMMA (Plexiglas) sheet where a grid of holes is drilled according to the mask rank. Thus, for Rank 7 the holes were 5.6mm in diameter and of 3mm pitch while for Rank 19 they were 2mm and 1mm, respectively. Each hole is filled with cylindrical blocks (Rank 7) or spheres (Rank 19) according to the MURA mask pattern. The size of the mask and the detector defines a solid angle, which is known as Fully Coded Field of View (FCFOV). A radioactive source within the FCFOV illuminates the mask and projects a fraction of its pattern (shadowgram) onto the detector surface. Since two detectors are utilized, a system FCFOV (denoted as $FCFOV_S$ hereafter) is introduced as the common area of the FCFOV of the detectors and is directly related to both their separation d and the mask-to-detector distance b. For Z>z_{min} the $FCFOV_S$ is 77.71×77.71 degrees.

The recorded shadowgram (figure 2) is the result of the convolution of the source with the mask as photons pass through the transparent and opaque elements [5],[6]. The applied method of deconvolution is by cross-correlating the shadowgram with a special function G, essentially a replica of the mask pattern itself. This function utilizes balanced correlation along with digital reconstruction to minimize artifacts [6]. The result of cross-correlation, denoted as *rec matrix*, defines the Source Point Spread Function (SPSF). Its peak corresponds to the trace of the source in pixel coordinates providing a signature of the source direction [7]. The displacement of the SPSF peak r_i ($SPSF_i$) from the center of the coordinate system (x_i, y_i, z_i) of the i_{th} detector (figure 1) designates the azimuthal angle (φ_i) while the elevation angle θ_i is calculated as θ_i=atan(r_i/b). This set of parameters (φ_i, θ_i)

Table 1. Main characteristics of the two experimental MURA masks

Rank	Number of elements	Diameter of elements (mm)	Surface (mm^2)	Grid pitch (mm)
7	13×13	5.6	72.8×72.8	3
19	37×37	2.0	74.0×74.0	1

specifies a unique direction that originates from the $SPSF_i$ and passes through the mask center. The triangulation algorithm gives the estimated coordinates of the radioactive source in the system reference frame (X, Y, Z). Given the set of parameters (φ_i, θ_i) from the reconstructed images, the algorithm finds the middle point of the minimum distance vector that is perpendicular to the two lines (figure 3).

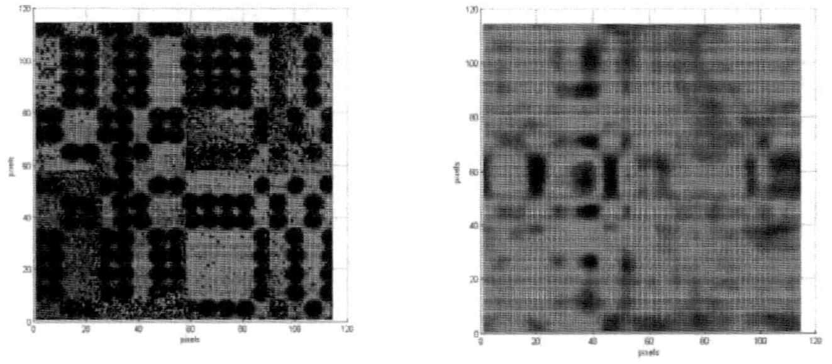

Figure 2. Recorded shadowgram (left) and the result of the deconvolution (rec matrix, right) with a clear signature of the SPSF (red spot).

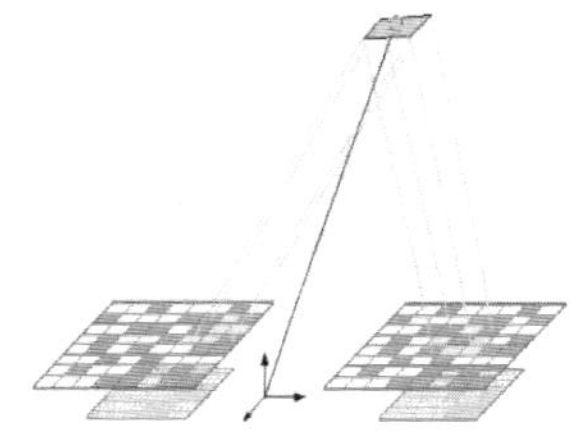

Figure 3. A schematic diagram that depicts the triangulation algorithm.

Results – discussion – conclusions

The Point Source Location Accuracy (PSLA) [8] and the Angular Resolution (AR) [9] are the figures of merit for the instrument. PSLA specifies how close the estimated location is to the actual position of the source while the AR defines the minimum angle where two sources can be clearly separated and identified. Both parameters have been estimated via Monte Carlo simulations.

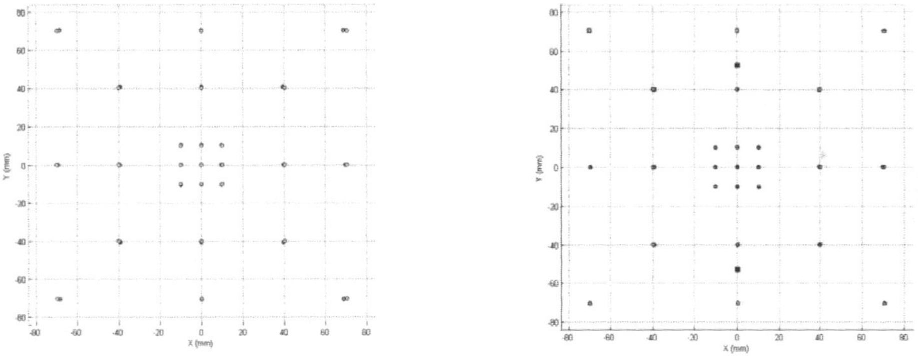

Figure 4. Source positions (yellow spots) projected in XY plane and their corresponding estimation (red boxes) at Z=150mm for Ranks 19 (left) and 7 (right), respectively. Blue squares indicate the actual position of the detectors.

In order to evaluate the PSLA with respect to MURA rank 7 and 19 masks, a ^{241}Am point-like source was positioned at several locations within the $FCFOV_S$ but at the same elevation Z=150mm (figure 4). The simulation experiment was iterated 50 times for each position. Table 2 shows some indicative results where the source location is estimated with high accuracy. With a mask to detector distance of 10mm, the AR of two point-like sources is 17.7° (figure 5) for Rank 7 and 12° for Rank 19. This gives minimum spacing between the two sources 48mm for the rank 7 and 32mm for the rank 19, at Z=150mm. When the size of the element of the rank 19 mask is changed to 1.85mm and the mask to detector distance becomes 18mm, then the angular resolution becomes 6.5° (figure 6), which at Z=110mm gives a minimum spacing of 12.5mm.

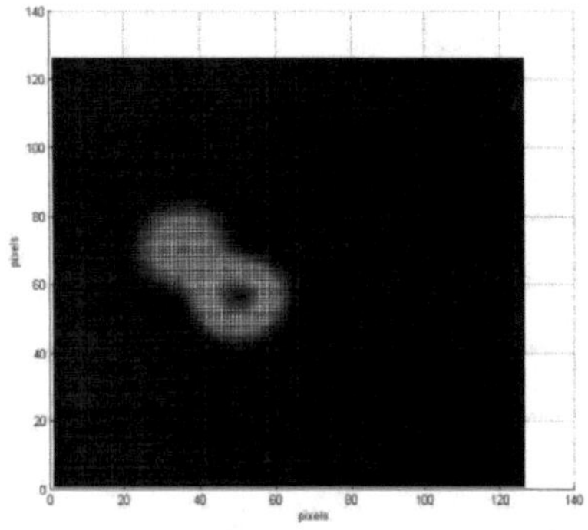

Figure 5. the result of reconstruction of two point like sources at Z=150mm with a mask of rank 7 and with element size 5.6mm

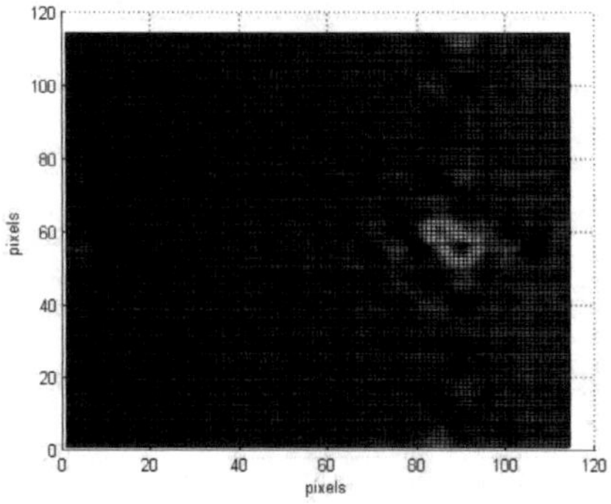

Figure 6. The result of reconstruction of two point like sources at Z=110mm with a mask of rank 19 and with element size 1.85mm

Triangulation gives the capability of 3-D localization, but it has no significant effect on the spatial resolution of the system, because the angular resolution of each mask-detector system determines the final system resolution. Further increase of the rank of the mask will result to higher spatial resolution.

Table 2. Indicative parameters for source location X=0, Y=0, Z=150mm.

Parameter	Rank 19	Rank 7	Unit
Total counts per detector	~33500	~37000	-
SNR per detector	34±2	31±2.7	-
X_{est} (@ 95% confidence level)	-0.09036 ± 0.3825	-0.0066 ± 0.4144	mm
Y_{est} (@ 95% confidence level)	0.06128 ± 0.3910	0.02180 ± 0.5101	mm
Z_{est} (@ 95% confidence level)	148.78 ± 0.6142	149.72 ± 1.7226	mm

Acknowledgement
This research has been funded by the ESA Contract 4200014240 «CdTe crystallization and related compounds» and by the FP7-SEC-218000 COCAE project.

References
[1] Tsuchimochi M., Hayama K. 2013 *Physica Medica* **29** 126
[2] Jeong S H, Baek H C, Son I Y, Cho Y D, Chung K M, Min Y J, Ko H Y and Kim T B 2006 *J. Korean Med. Sci.* **21** 865
[3] F Hirofumi, Idoine D J, Gioux S, Accorsi R, Slochower R D, Lanza C R and Frangioni V J 2012 *Mol. Imaging Biol* **14** 173
[4] Meng L.J., Tan J.W., Spartiotis K., Schulman T. 2009, *Nucl. Instr. Meth. Phys. Res. A* **604** 548
[5] Anderson N D, Stromswold C D, Wunschel C S, Peurrung J A and Hansen R R 2006, *Technometrics* **48**, no. 2 252
[6] Lees E J, Bugby L S, Bassford J D, Blackshaw E P and Perkins C A 2013, *J. Inst.* **8** 10021
[7] Fenimore E E and T. M. Cannon M T 1981 *Applied Optics* **20** 10 1858
[8] Gmar M., Agelou M., Carrel F., Schoepff V. 2011 *Nucl. Instr. Meth. Phys. Res. A* **652** 638
[9] Caroli E, Stephen B J, Cocco D G, Natalucci L and Spizzichino A 1987 *Space Sci. Rev.* **45** 349
[10] Gros A, Goldwurm A, Cadolle-Bel M, Goldoni P, Rodriguez J, Foschini L, Santo D M and Blay P 2003 *Astronomy & Astrophysics* **411** 179

Contrast to Noise Ratio and Contrast Detail Analysis in Mammography: A Monte Carlo Study

V Metaxas[1], H Delis[1], C Kalogeropoulou[2], P Zampakis[2], G Panayiotakis[1]

[1]Department of Medical Physics, School of Medicine, University of Patras, Patras, Greece.
[2]Department of Radiology, School of Medicine, University of Patras, Patras, Greece.

E-mail: panayiot@upatras.gr

Abstract. The mammographic spectrum is one of the major factors affecting image quality in mammography. In this study, a Monte Carlo (MC) simulation model was used to evaluate image quality characteristics of various mammographic spectra. The anode/filter combinations evaluated, were those traditionally used in mammography, for tube voltages between 26 and 30 kVp. The imaging performance was investigated in terms of Contrast to Noise Ratio (CNR) and Contrast Detail (CD) analysis, by involving human observers, utilizing a mathematical CD phantom. Soft spectra provided the best characteristics in terms of both CNR and CD scores, while tube voltage had a limited effect. W-anode spectra filtered with k-edge filters demonstrated an improved performance, that sometimes was better compared to softer x-ray spectra, produced by Mo or Rh anode. Regarding the filter material, k-edge filters showed superior performance compared to Al filters.

1. Introduction

The carcinogenic risk associated with the exposure of the breast tissue, in addition to the requirements for high image quality have made optimization of mammography an issue of high importance [1-4]. Technological developments have led to Digital Mammography (DM). DM can be combined with post-processing algorithms, to compensate a possibly suboptimal image contrast and improve lesion detectability. This is mainly due to the separation of image acquisition and display processes, that enables independent optimization of each one. However, any contrast amplification is limited by the subject contrast, because as the displayed contrast is increased, so is the visibility of noise.

A critical factor determining image quality is the mammographic spectrum, characterized by anode/filter material combination and tube voltage. The detectability of low-contrast breast masses is most limited by low-contrast resolution [5]. An objective image quality metric should incorporate the measurement of the CNR in selected Regions of Interest (ROIs). Image quality may also be characterized by subjective methods, utilizing CD phantoms [6]. These phantoms provide a tool for conducting CD experiments by individuating the boundary between visible and invisible objects [7].

Several experimental and simulation studies have been carried out, in order to investigate the effect of mammographic spectrum on image quality. In Screen-Film (SF) mammography, Mo/Mo combination has been considered the optimal choice, especially for thinner breasts, but other choices have also been introduced [8-10]. For DM, W-anode spectra, combined with proper k-edge filtration, provide improvement of image quality, especially for thicker breasts [1-5, 10-15].

In this study, a validated MC simulation model dedicated for mammography [16-18] was used with a mathematical CD phantom, to perform quantitative and qualitative evaluation of the imaging performance of several mammographic spectra, in terms of CNR and CD analysis.

2. Materials and methods

2.1. Simulation of mammographic procedure

The MC model is described in detail in previous works [16-18]. The exposure of the photons exiting the phantom towards the detector plane was recorded and the two dimensional exposure map (with pixel size of 25 μm) was transformed to an 8-bit greyscale image, to illustrate the results in a pictorial form (subject contrast). In this study, the effect of detector was not considered. In order the results to be comparable, the number of the MC photons was selected to produce the same exposure just above the detector plane. This exposure was selected to be 200 mR, in order to balance between acceptable statistical fluctuation and reasonable simulation computational time. MC experiments were performed for 396 different cases combining various sizes of inhomogeneities, anode/filter combinations and tube voltages (Table 1). The mean execution time was 2h.

2.2. Geometrical model
The main geometrical model adopted, was the one presented by Spyrou et al. [16-18]. The focal spot dimensions were considered to be 0.01 x 0.01 cm^2, while the x-ray intensity distribution was selected to be uniform. The collimators were placed 15 cm from the focus, corresponding to a Field of View (FOV) of 0.42 x 0.42 cm^2 on the detector plane. This specific FOV was selected in order to irradiate only the appropriate part of the phantom and therefore to reduce the executional time. The antiscatter grid was removed, since the selected FOV offers, to some extend, limitation of inherent scatter.

The breast phantom was specifically designed, to allow CD analysis. The compressed breast was represented by a semi-cylinder with 4 cm radius, positioned to an x-ray focus-to-detector distance of 60 cm. The background material was selected to be PMMA with a thickness of 1 cm. The phantom included two identical spherical air inhomogeneities of logarithmically varied diameters (90-250 μm) (Table 1), in order to simulate low-contrast details. The first inhomogeneity was placed at the centre of the phantom and the second was randomly placed at one of the four corners of the phantom.

2.3. Mammographic spectra
The mammographic spectra utilized were computed using a previously developed and validated analytical method [2]. Twelve anode/filter combinations, with 1mm Beryllium (Be) inherent filtration, were considered for tube voltages between 26 and 30 kVp. The anode/filter combinations (Table 1), were based on combinations proposed, either for DM [1-5, 10-15], or for SF technique [8-9].

Table 1. Anode/filter material, tube voltage and inhomogeneity sizes studied. The thickness of each filter is indicated in parenthesis.

Anode	Filter	Tube Voltage (kVp)	Inhomogeneity Size (μm)
Mo	Mo (30 μm), Nb (30 μm), Rh (29 μm), Al (510 μm)	26, 28, 30	90-250
Rh	Rh (29 μm), Al (510 μm)		
W	Mo (30 μm), Nb (50 μm), Rh (30 μm), Pd (50 μm)+2mm Al, Zr (50 μm), Al (500 μm)		

2.4. Quantitative evaluation of image quality
On the generated Monte Carlo images, CNR was derived according to the formula [4]:

$$CNR = \frac{|C_{INH} - C_{BG}|}{\sqrt{N^2_{INH} + N^2_{BG}}}$$

where C_{INH} and N_{INH} is the mean grey level value and its standard deviation in the selected ROI under the inhomogeneity, and C_{BG} and N_{BG} is the mean grey level value and its standard deviation at the same point with the absence of the inhomogeneity. CNR values were normalized to the case of Mo/Mo anode/filter material combination at 28 kVp. The generated images were analysed using ImageJ [19].

2.5. Observer study
A four-alternative experiment was conducted, as a result of the randomly positioned inhomogeneity. For verifying the detection of each inhomogeneity, the observers were asked to indicate the corner where the inhomogeneity was located. Images were evaluated on a dedicated high resolution monitor.

The images were presented in a random order, with the room light off, while the observers were free to adjust magnification factor, brightness (window and level settings) and contrast, in order to get the best visibility. Two experienced observers independently reviewed all the images in one session. 396 images were evaluated plus an additional sample of 100 random copies, for the evaluation of the intraobserver variability. Observer's performance was based on a CD index, related to the visibility of the randomly positioned inhomogeneity. A 0-3 scale of visibility was used [20]. The individual visibility scores were averaged to obtain the mean CD score. The CD index was defined as the product of the CNR of each inhomogeneity with the corresponding mean CD score. Statistical analysis was performed using the SPSS statistical package. A p-value of less than 0.05 was considered to indicate the threshold of statistical significance.

3. Results and Discussion

3.1. Contrast to Noise Ratio

The CNR of small low-contrast details strongly depends on mammographic spectrum. Beam hardening results in a decrease of signal differences between the inhomogeneity and the background and thus the subject contrast is degraded. Grey level values, for the case of Mo/Mo spectrum, are presented in Figure 1a. The corresponding values of CNR are presented in figure 1b, for all inhomogeneity sizes. The CNR decreases with the inhomogeneity size, while an increase of tube voltage leads to CNR decrease. The reduction of CNR is due to the fact that beam hardening decreases noise, but at the same time significantly decreases signal differences between inhomogeneity and background (figure 1a), since Compton scattering is the most dominant effect at these energies. The same trend appeared for all spectra, as observed in figure 2a, which presents the CNR values obtained with a 110 μm inhomogeneity. Increasing tube voltage from 26 to 30 kVp, the CNR decreases up to 22% for the 110 μm and 13% for the 250 μm detail.

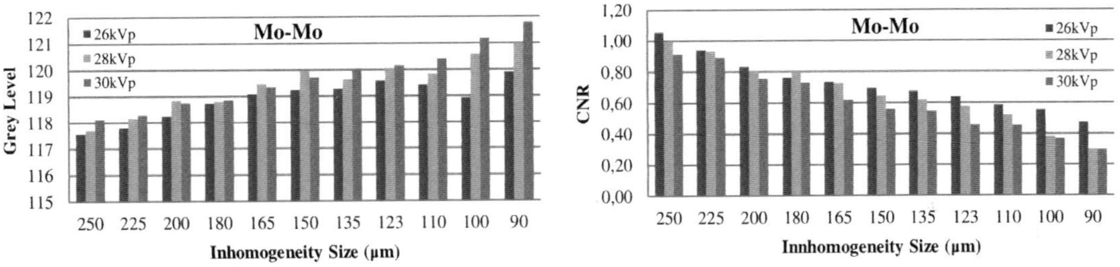

Figure 1. a) Signal strength, and **b)** CNR for all inhomogeneity sizes and tube voltages studied, for the case of Mo/Mo anode/filter combination.

The influence of anode/filter material combination on the CNR is presented in figure 2a. Concerning filter material, k-edge filters demonstrated improved performance compared to Al filter, especially for smaller inhomogeneities. In the case of Rh-anode spectra, if an Al-filter is used instead of Rh-filter, the CNR is reduced up to 23% for the 110 μm inhomogeneity. In general, as the size of inhomogeneity decreases, softer beams are required to achieve better CNR. The performance of k-edge filters is reversely proportional to the energy of their absorption edge. These filters cut off the photons with energy above their absorption edge, which causes subject contrast degradation due to the increased penetration through the breast. Regarding anode material, W- anode spectra demonstrated similar performance, which sometimes was better compared to softer x- ray spectra, produced by Mo or Rh anodes. Furthermore, W/0.050mmPd+2.00mmAl combination demonstrated the worst CNR.

3.2. Contrast-Detail Analysis

The main objective of CD experiments was to investigate the detection of inhomogeneities. No significant differences were found between and within the two observers. Figure 2b presents the influence of mammographic spectrum on the observers' performance. For most anode/filter material combinations, the highest CD scores occurred at the lowest tube voltage and tend to decrease as the tube voltage increases. Indicatively, the relative CD score differences are approximately 18% for the case of Mo/Mo between 26 and 30 kVp. It is interesting to notice that CD scores tend to increase as a function of kVp only for W/Mo anode/filter combination, possibly due to noise reduction [14]. Regarding filter material, k-edge filters demonstrated a superior performance compared to Al filters, with their performance being reversely proportional to the energy of their absorption edge. The latter is more obvious in the case of Mo-anode spectra. This relationship is not monotonous for W-anode spectra as noise reduction associated to harder spectra results to the enhancement of the CNR and consequently to improved detectability of inhomogeneities (figures 2a, 2b) [14]. W-anode spectra showed an improved performance, which sometimes was better or comparable to softer x-ray spectra produced by Mo- or Rh-anodes, especially when combined with k-edge filters of lower k-absorption edge. It should be noted, that low-contrast detection was investigated over a uniform CD phantom. These conditions do not simulate the clinical practice, since apart from the quantum noise, additional structure noise limits the detection of details.

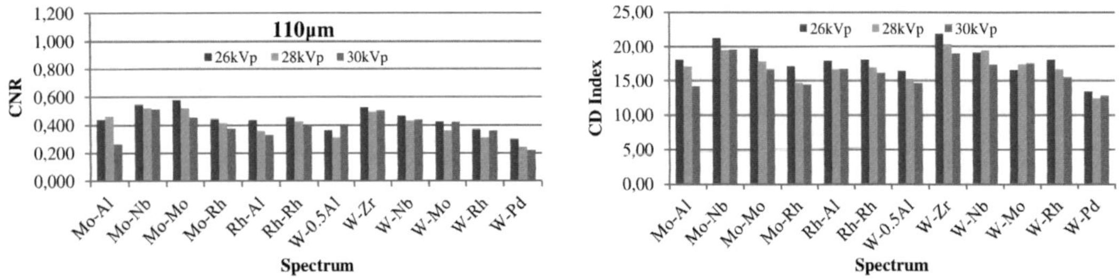

Figure 2. a) CNR for the various spectra studied, for a 110 μm detail, **b)** CD scores for the various spectra studied.

4. Conclusion

The detection of small low-contrast details was enhanced by using a softer x-ray beam, when the exposure and thickness of the breast tissue remain constant. Optimum performance, in terms of CNR and CD scores, was achieved by selecting the lowest tube voltage value in combination with filters of low energy k-edge.

References

[1] Delis H et al 2005 *Eur. J. Radiol.* **54** 371-376.
[2] Delis H et al 2006 *Med. Phys.* **33** 4221–4235.
[3] Delis H et al 2007 *Nucl. Instrum. Meth. A* **580** 493–496.
[4] Koutalonis et al 2007 *Phys. Med. Biol.* **52** 3185-3199.
[5] Williams M B et al 2008 *Med. Phys.* **35** 2414–2423.
[6] Rivetti S et al 2006 *Med. Phys.* **33** 4198-4209.
[7] Bijkerk K R et al 2000 Manual CDMAM phantom type 3.4 (Nijmegen, Netherlands).
[8] Gingold L E et al 1995. *Radiol.* **195** 639-644.
[9] Thilander-Klang A C et al 1997 *Radiol.* **203** 348-354.
[10] Dance D R et al 2000 *Br. J. Radiol.* **73** 1056–1067.
[11] Bernhardt P et al 2006 *Med. Phys.* **33** 4337–4349.
[12] Tomal A et al 2010 *Appl. Radiat. Isotop.* **68** 755–759.
[13] Cunha D M et al 2012 *Phys. Med. Biol.* **57** 1919–1935.
[14] Berns E A et al 2003 *Med. Phys.* **30** 334–340.
[15] Baldelli P et al 2010 *Br. J. Radiol.* **83** 290–295.
[16] Spyrou G et al 1998 *Phys. Med. Biol.* **43** 3341-3357.
[17] Spyrou G et al 2000 *Med. Infor.* **25** 275-293.
[18] Spyrou G et al 2002 *Phys. Med. Biol.* **47** 917-933.
[19] http://rsb.info.nih.gov/ij/.
[20] Undrill et al 2000 *Clin. Radiol.* **55** 782-790.

Image Quality Assessment of a CMOS/Gd_2O_2S:Pr,Ce,F X-ray Sensor

C. M. Michail[1], I. E. Seferis[2], T. Sideras[1], I. G. Valais[1], G. P. Fountos[1], A. Bakas[3], G. S. Panayiotakis[4] and I. S. Kandarakis[1]

[1]Radiation Physics, Materials Technology and Biomedical Imaging Laboratory, Department of Biomedical Engineering, Technological Educational Institute of Athens, Egaleo, 122 10 Athens, Greece

[2]Faculty of Chemistry, Wroclaw University, 14F Joliot-Curie Street, 50-383 Wroclaw, Poland

[3]Medical Radiological Technology, Faculty of Health and Caring Professions, Technological Educational Institute of Athens, 122 10 Athens, Greece

[4]Department of Medical Physics, Faculty of Medicine, University of Patras, 265 00 Patras, Greece

e-mail: michail@upatras.gr

Abstract. The aim of the present study was to examine the image quality performance of a CMOS digital imaging optical sensor coupled to custom made gadolinium oxysulfide powder scintillators, doped with praseodymium, cerium and fluorine (Gd_2O_2S:Pr,Ce,F) screens. The screens, with coating thicknesses 35.7 and 71.2 mg/cm^2, were prepared in our laboratory from Gd_2O_2S:Pr,Ce,F powder (Phosphor Technology, Ltd) by sedimentation on silica substrates and were placed in direct contact with the optical sensor. Image quality was determined through a single index image quality parameter (information capacity). The CMOS sensor/Gd_2O_2S:Pr,Ce,F screens combinations were irradiated under the RQA-5 (IEC 62220-1) beam quality. The detector response function was linear for the exposure range under investigation. Under the general radiography conditions, both Gd_2O_2S:Pr,Ce,F screen/CMOS combinations exhibited comparable overall imaging properties, in terms of the information capacity, to previously published scintillators, such as Gd_2O_2S:Eu.

Keywords: Information Capacity; CMOS; Inorganic Scintillators; Gd_2O_2S:Pr,Ce,F

1. Introduction

Indirect medical imaging detectors incorporate a scintillator detector coupled to an optical sensor (CCD, CMOS, etc.). In this case, the scintillator is used in the form of a powder phosphor screen [1]. Optical ceramic scintillators have been also developed in order to replace single crystals in some applications [1-4]. Ceramic gadolinium oxysulfide doped with praseodymium, cerium and fluorine (Gd_2O_2S:Pr,Ce,F) (ultra-fast ceramics-UFC), has been proposed to be used in Computed Tomography (CT) scanners [4]. Gd_2O_2S:Pr,Ce,F is an efficient and fast scintillator (decay time of the order of 3-4 μs) based on the well-known traditional Gd_2O_2S host material [5].

A semiconductor technology that has been used widely recently is complementary metal oxide (CMOS) semiconductors [6-8]. Active pixel sensor (APS) CMOS provide high resolution even at high framing rates and, in association with scintillating screens, have been increasingly investigated for medical imaging applications [9]. In the present study, image quality of a CMOS digital imaging sensor, coupled to Gd_2O_2S:Pr,Ce,F powder phosphor screens, was investigated by assessing the information content through the information capacity (IC) [10].

2. Materials and Methods

2.1. Phosphor screens

Gd_2O_2S:Pr,Ce,F was purchased in powder form (Phosphor Technology Ltd, England, code: UKL59CF/N-R1) with a mean grain size (estimated by ultrasonic dispersion with a coulter counter having 100 μm aperture) of approximately 16.0 μm at the 95% of the volume and quartile deviation of 0.25 (Phosphor Technology Ltd., datasheet) [5]. Gd_2O_2S:Pr,Ce,F has effective atomic number $Z_{eff} = 61.1$, density of 7.34 g/cm^3 and a decay time of the order of a $3X10^{-6}$ s [3]. The phosphor was used in the form of thin layers to simulate the intensifying screens employed in X-ray imaging [11]. Two screens with coating thicknesses 35.7 and 71.2 mg/cm^2 were prepared by sedimentation of Gd_2O_2S:Pr,Ce,F powder on fused silica substrates (spectrosil B). Sodium orthosilicate (Na_2SiO_3) was used as binding material between the powder grains [12].

2.2. CMOS sensor

The Gd_2O_2S:Pr,Ce,F scintillating screens were manually coupled to an optical readout device including a CMOS Remote RadEye HR photodiode pixel array [13]. The CMOS photodiode array consists of 1200x1600 pixels with 22.5 μm pixel spacing. The Gd_2O_2S:Pr,Ce,F screens were directly overlaid onto the active area of the CMOS photodiode array, consisting of an N-well diffusion on p-type epitaxial Silicon. A 70 kV (RQA-5) X-ray beam was used, following the IEC standards [14]. IEC standard X-ray spectrum was achieved by adding 21 mm Al filtration in the beam to simulate beam quality alternation by a human body [15]. A BMI General Medical Merate tube with rotating Tungsten anode and inherent filtration equivalent to 2 mm Al, was used for the RQA-5 beam quality. According to IEC standard, the source-to-detector distance (SDD) between the X-ray focal spot and the surface of the detector should be no less than 150cm. In this study was set to 176 cm. The added filtration was placed as close as possible to the source.

2.3. Image quality

2.3.1. Signal Transfer Property (STP)

The Signal Transfer Property (STP) provides the relationship between mean pixel value (MPV) and incident air Kerma (ESAK) at the detector surface. This relationship was obtained by plotting pixel values versus ESAK at the detector, as described in the IEC method [14]. A sequence of uniform images was acquired at different exposure levels. MPV was evaluated in a 1×1 cm^2 region of interest (ROI). The system's response curve was fitted using a linear equation of the form $MPV = \alpha + b \times K_\alpha$, where α and b are fit parameters and K_α the ESAK in units of μGy. From the slope of the system's response curve, the value of the gain factor (G) was obtained [16]. The magnitude of the pixel offset at zero air-Kerma was also estimated [17].

2.3.2. Information Capacity (IC)

The concept of image information capacity (IC) has been introduced within the context of Shannon's information theory, in order to assess image information content [18-23]. In digital imaging, the continuous spatial distribution of an optically generated image is sampled by the discrete sensitive pixels on a photodiode array, whose outputs are converted into digitized signals and stored in an image processing system for numerical evaluation. In this study information capacity was calculated according to Equation (1) [24]:

$$IC = 1/2 \int_0^u \log_2(1 + SPS(u)/NPS(u)) s\pi u du \qquad (1)$$

where SPS is the signal power spectrum, defined as: $SPS = (MTF \times G)^2$ [24].

3. Results and Discussion

Figure 1 shows the detector response curves (STP) of the CMOS sensor combined with the 35.7 and 71.2 mg/cm^2 Gd_2O_2S:Pr,Ce,F screens, respectively under the RQA-5 (70kVp) beam quality. The detector was found to have a linear response, covering the whole exposure range, with a pixel value offset of -9.046 and 16.63 for the 35.7 and 71.2 mg/cm^2 Gd_2O_2S:Pr,Ce,F screens. The linear no threshold fits gave correlation coefficients (R^2) greater than 0.9987 and 0.9989 for the 35.7 and 71.2

mg/cm^2 Gd_2O_2S:Pr,Ce,F screens. Using flat-field images the gain factors were determined by linear regression to be G=2.610 and 4.379 digital units per μGy for the 35.7 and 71.2 mg/cm^2 Gd_2O_2S:Pr,Ce,F screens.

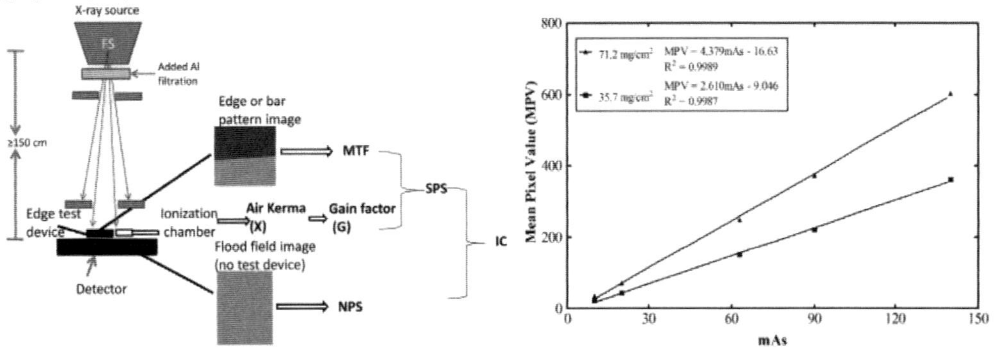

Figure 1. Experimental setup (left) and detector response curves (STP) of the CMOS sensor combined with the 35.7 and 71.2 mg/cm^2 Gd_2O_2S:Pr,Ce,F screens, respectively under the RQA-5 (70kVp) beam quality (right).

Table 1 shows information capacity values for the combination of the CMOS sensor with the 35.7 and 71.2 mg/cm^2 Gd_2O_2S:Pr,Ce,F screens under investigation, and previously published IC values for a CMOS sensor coupled to Gd_2O_2S powder scintillators, activated either with Terbium (Tb) or Europium (Eu). The investigation was carried out under the RQA-5 X-ray spectrum, as a function of air-Kerma. The thinner Gd_2O_2S:Tb screen (33.91 mg/cm^2) showed the highest IC values due to the screen thickness and the higher MTF values of this screen, compared to the 35.7 mg/cm^2 Gd_2O_2S:Pr,Ce,F coupled to the CMOS sensor. The IC values of the thicker 71.2 mg/cm^2 Gd_2O_2S:Pr,Ce,F were lower than all the other screens, even from the IC value of the 65.1 mg/cm^2 Gd_2O_2S:Eu screen due to the fact of the higher thickness and screen uniformity, leading to lower MTF values and higher noise. The comparison was obtained at the same exposure level for all screen/sensor combinations. These data show that, for a given level of incident X-ray fluence, information capacity is mainly determined by the intrinsic phosphor material properties and by the screen thickness of the imaging system. In thick screens the lateral light trajectories are very long causing a large fraction of the laterally directed photons to be absorbed before reaching the screen output.

Table 1. Information Capacity values.

		Information capacity (bits/mm^2)			
Beam quality	Coating weight (mg/cm^2)	CMOS-Scintillator combinations			
		Gd_2O_2S:Tb	Gd_2O_2S:Eu	Gd_2O_2S:Pr,Ce,F	MTF @10% (lp/mm)
RQA-5	33.91	2806±34	-	-	9.1
	65.1	-	1826±18	-	5.0
	35.7	-	-	1813±23	6.7
	71.2	-	-	1722±35	4.8

4. Conclusions

In the present study image quality of two custom made Gd_2O_2S:Pr,Ce,F powder scintillator screens, coupled to a CMOS digital imaging sensor, was investigated under X-ray radiography imaging conditions. Image quality was investigated in terms of the information capacity, in the general radiography energy range. The detector response function was linear for the exposure range under investigation. The overall imaging properties, in terms of IC, of both Gd_2O_2S:Pr,Ce,F screen/CMOS combinations, can be considered for medical imaging applications since it was found comparable with previously published scintillators, such as the Gd_2O_2S:Eu.

Acknowledgements
Authors wish to thank Dr. Nektarios Kalyvas for his contribution in noise assessment.

5. References

[1] Nikl M, 2006 *Meas. Sci. Technol.* **17** R37.
[2] Yamada H, Suzuki A, Uccida Y, Yoschida M, Yammoto H 1989 *J. Electrochem. Soc.* **136** 2713.
[3] Greskovich C and Duclos S 1997 *Annu. Rev. Mater. Res.* **27** 69.
[4] Veronese I Radiation physics for Nuclear Medicine. Springer, Heidelberg, 166, 2011.
[5] Michail C, Valais I, Seferis I, Kalyvas N, David S, Fountos G and Kandarakis I 2014 *Radiat. Meas.* **70** 59.
[6] Bohndiek S, Cook E, Arvanitis C, Olivio A, Royle G, Clark A, Prydderch M, Turchetta R, Speller R 2008 *Phys.Med. Biol.* **53** 655.
[7] Seferis I, Michail C, Valais I, Zeler J, Liaparinos P, Fountos G, Kalyvas N, David S, Stromatia F, Zych E, Kandarakis I, Panayiotakis G 2014 *J. Lumin.* **151** 229.
[8] Michail C, Valais I, Seferis I, Kalyvas N, Fountos G and Kandarakis I 2015 *Radiat. Meas.* **74** 39.
[9] Endrizzi M, Oliva P, Golosio B, Delogu P 2013 *Nucl. Instrum. Meth. A* **703** 26.
[10] Michail C, Kalyvas N, Valais I, Fudos I, Fountos G, Dimitropoulos N, Koulouras G, Kandris D, Samarakou M and Kandarakis I 2014 *Biomed. Res. Int.* **2014**, 634856.
[11] Michail C, Kalyvas N, Valais I, David S, Seferis I, Toutountzis A, Karabotsos A, Liaparinos P, Fountos G and Kandarakis I 2013 *J. Lumin.* **144** 45.
[12] Michail C, Fountos G, Valais I, Kalyvas N, Liaparinos P, Kandarakis I, Panayiotakis G 2011 *IEEE Trans. Nucl. Sci.* **58(5)** 2503.
[13] Michail C, Spyropoulou V, Fountos G, Kalyvas N, Valais I, Kandarakis I and Panayiotakis G 2011 *IEEE Trans. Nucl. Sci.* **58(1)** 314.
[14] Medical Electrical Equipment-Characteristics of Digital X-Ray Imaging Devices, IEC, Switzerland, IEC 62220-1-2 2005.
[15] Michail C, David S, Liaparinos P, Valais I, Nikolopoulos D, Toutountzis A, Cavouras D, Kandarakis I, Panayiotakis G 2007 *Nucl. Instrum. Meth. A* **580** 558.
[16] Neitzel U, Gunther-Kohfahl S, Borasi G, Samei E 2004 *Med. Phys.* **31** 2205.
[17] Samei E, Flynn M and Reimann D 1998 *Med. Phys.* **25** 102.
[18] Shannon C 1948 *Bell Syst. Tech. J.* **27** 379.
[19] Jones R 1961 *J. Opt. Soc. Am.* **50** 1166.
[20] Jones R 1962 *J. Opt. Soc. Am.* **52** 1193.
[21] Kanamori H 1968 *Japan. J. Appl. Phys.* **7** 414.
[22] Kanamori H and Matsuoto M 1984 *Phys. Med. Biol.* **29** 303.
[23] Wagner R, Brown D and Paster M 1979 *Med. Phys.* **6** 83.
[24] Seferis I, Michail C, Valais I, Fountos G, Kalyvas N, Stromatia F, Oikonomou G, Kandarakis I, Panayiotakis G 2013 *Nucl. Instrum. Meth. A* **729** 307.

Assessment of the Contrast to Noise Ratio in PET Scanners with Monte Carlo Methods

C. M. Michail[1], G. E. Karpetas[2], G. P. Fountos[1], I. G. Valais[1], D. Nikolopoulos[3], I. S. Kandarakis[1] and G. S. Panayiotakis[2]
[1]Radiation Physics, Materials Technology and Biomedical Imaging Laboratory, Department of Biomedical Engineering, Technological Educational Institute of Athens, Egaleo, 122 10 Athens, Greece
[2] Department of Medical Physics, Faculty of Medicine, University of Patras, 265 00 Patras, Greece
[3] Department of Computer Electronic Engineering, Technological Education Institute of Piraeus, Greece, Petrou Ralli & Thivon 250, 122 44, Aigaleo, Athens, Greece

e-mail: gfoun@teiath.gr

Abstract. The aim of the present study was to assess the contrast to noise ratio (CNR) of PET scanners through a thin layer chromatography (TLC) plane source. The source was simulated using a previously validated Monte Carlo model. The model was developed by using the GATE MC package and reconstructed images obtained with the STIR software for tomographic image reconstruction. The PET scanner simulated was the GE DiscoveryST. A plane source consisted of a TLC plate, was simulated by a layer of silica gel on aluminum (Al) foil substrates, immersed in 18F-FDG bath solution. Image quality was assessed in terms of the CNR. CNR was estimated from coronal reconstructed images of the plane source. Images were reconstructed by the maximum likelihood estimation (MLE)-OSMAPOSL. OSMAPOSL reconstruction was assessed by using various subsets (3, 15 and 21) and various iterations (2 to 20). CNR values were found to decrease when both iterations and subsets increase. Two (2) iterations were found to be optimal. The simulated PET evaluation method, based on the TLC plane source, can be useful in image quality assessment of PET scanners.

Keywords: PET; CNR; image quality; Monte Carlo

1. Introduction

In positron emission tomography (PET) scanners, iterative reconstruction, based on maximum likelihoods such as the ordered subset expectation maximization (OSEM) algorithm, has been widely adopted for clinical practice [1]. However, in OSEM, resolution and noise cannot be assessed in a straight manner, such as in filtered backprojection reconstructed images, since iterative reconstruction algorithms are non-linear and difficult to evaluate [2]. The major consideration, when using iterative reconstruction algorithms, is to retain noise and processing time into acceptable levels. To this aim, the number of iterations should be carefully selected in order to compromise between resolution and noise-processing time [3]. In clinical practice, where the number of iterations is limited due to time restrictions, resolution may differ in different objects within an image, depending on their relative size and intensity [2]. The number of iterations affects both the resolution and intensity of PET in the following manner. By increasing the number of iterations, within a reasonably wide range of practically encountered clinical situations [1], resolution improves and tumors appear brighter. However, this behavior reaches to a point that thereafter resolution cannot be further improved [4,5]. The aim of this study was to assess both resolution and noise, in terms of the contrast to noise ratio (CNR) [6,7]. Increased CNR values preserve spatial resolution, while retaining noise into acceptable levels and in a clinical context, could allow better detection and earlier diagnosis [6]. To this aim coronal images of a flood source were obtained [8] following the work of Fountos *et al.* [9,10] for SPECT scanners resolution assessment. A previously validated Monte Carlo model [11,12] was used in order to obtain the PET images, by using different number of iterations and subsets in the iterative image reconstruction. Within the context of the model, CNR was estimated by the simulation of a thin layer chromatography (TLC) plane source filled with 18F-FDG using a software model based on the GATE Monte Carlo package. The GATE package is an open-source extension of the Geant4 Monte Carlo toolkit [13]. GATE was used in combination with the STIR image reconstruction software [14] to obtain the plane source reconstructed images. The simulation of this plane source phantom provides an accurate model that is useful to fully characterize the performance of nuclear medicine imaging systems.

2. Materials And Methods

1.1 Geometry of the modeled PET scanner

In this study the Discovery ST PET/CT system was modeled. The system incorporates bismuth germinate oxide (BGO) [15,16] crystals with dimensions of 6.3x6.3x30 mm^3 in the tangential, axial and radial directions, respectively. The detector ring is finally comprised of 35 modules, i.e. 280 crystal blocks, or 24 rings of 420 crystals (for a total of 10080 BGO crystals). The dimensions of the rings are 88.6 cm diameter with a 15.7 cm axial and 70cm transaxial fields of view (FOV).

2.2. Software for simulation

The simulation software, employed in the present study, was built around the GATE toolkit, which in turn is based on the Geant4 platform. In every BGO crystal, of the detector block, a Gaussian energy blur is applied with an average energy resolution of 17%, referenced at 511 keV. Afterwards, a 300 ns dead time value was applied on the single events in the BGO crystal [17] by a paralysable Deadtime module. The Quantum Detection Efficiency (QDE) [18,19] of the crystals was found 0.94 at 511 keV. The sinogram output file (.ima) was used by STIR as input file for the reconstruction of the simulated plane source image. The coincidence time window was set to 11.7 ns.

2.3. Preparation of the phantom

The phantom was a planar source similar to those used in previous works of Boone [20] for CT and Fountos *et al.* [9,10] for SPECT systems. The idea for the particular 18F-FDG plane source was based on the excellent binding of 18F-FDG with TLC plates [8]. The plate was considered to consist of a layer of silica gel on Al foil substrates (Al density 2.7 g/cm^3). The dimensions of the TLC plate were 5x10 cm^2 and it was assumed to be immersed in an 18F-FDG bath solution (1 MBq). The plane source (i.e. the radioactive plate) was simulated between two semi-cylindrical polyethylene blocks with 20 cm diameter and 70 cm length, in the horizontal and vertical direction.

2.4 Contrast to Noise Ratio (CNR)

CNR was obtained, by using equation (1) [6,7].

$$CNR = \frac{\overline{\mu_2} - \overline{\mu_1}}{\sqrt{SD_1{}^2 + SD_2{}^2}} \tag{1}$$

Where μ_1 and μ_2 are the mean pixel values, calculated from regions of interest (ROIs) within hot (in the source) and cold (out of the source) areas respectively. SD_1, SD_2 are the corresponding standard deviation (SD) values.

3. Results and Discussion

3.2 Contrast to Noise Ratio (CNR)

Figure 1 shows the resulted CNR values when the number of iterations and subsets are allowed to vary, in order to investigate the impact on PET image quality. The examined combinations of iterations and subsets are:3 subsets with 2, 6,8,14 and 20 iterations, 15 subsets with 4, 6,12,14 and 20 iterations, and 21 subsets with 2, 6,10,14 and 20 iterations. As it can be seen, CNR values decrease when the number of iterations and subsets increase. When the number of iterations increases, contrast also improves, but with a corresponding increase in image noise. However the increase in image noise is of greater dimensions, compared with resolution improvement, leading to a decrease of the CNR values. Optimal CNR was obtained for 2 iterations [6] and 3 subsets.

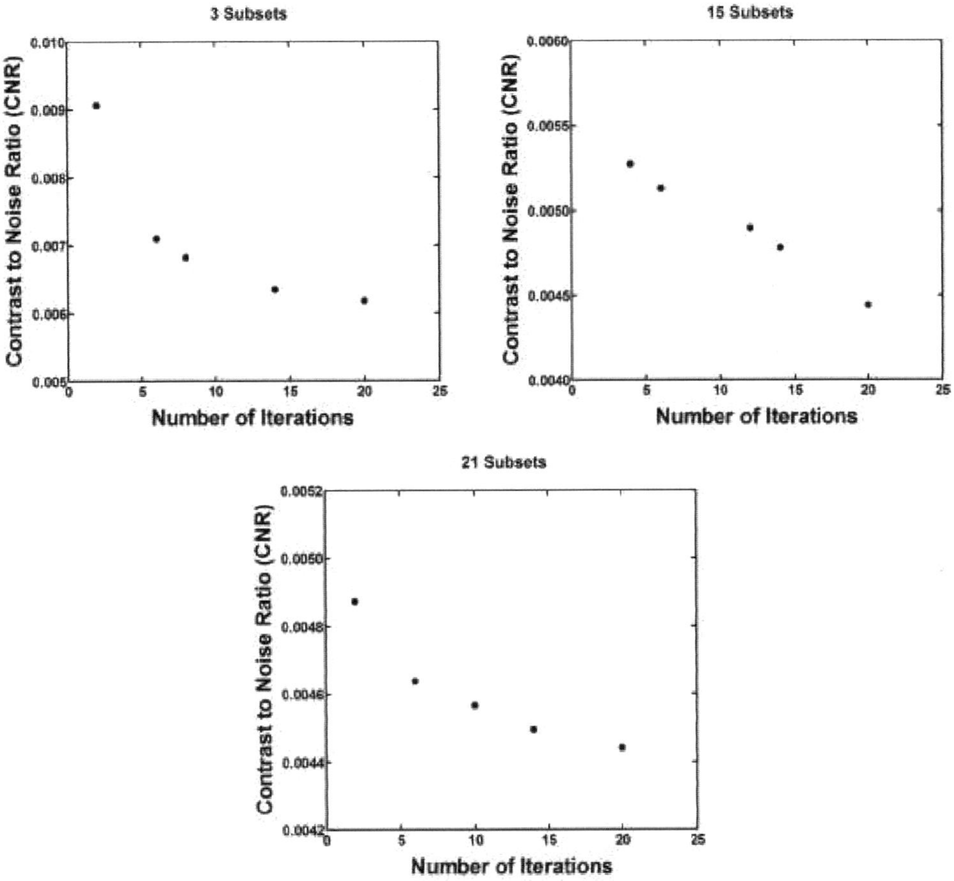

Figure 1. CNR values obtained from the OSMAPOSL reconstructed images using various iterations and subsets a) 3 subsets, 2, 6,8,14 and 20 iterations, b) 15 subsets, 4, 6,12,14 and 20 iterations, c) 21 subsets, 2, 6,10,14 and 20 iterations.

4. Conclusions

Image quality characterisation of a PET scanner was achieved through the estimation of the CNR of thin 18F-FDG plane source reconstructed images, simulated by a Monte Carlo model. The influence of iterative image reconstruction on the plane source simulated images was investigated by using various subsets and iterations. CNR values were found to decrease by increasing the number of iterations (number of image updates) and subsets. The method was used for the image quality assessment and optimisation, but it can be also useful in research for the further development of PET and SPECT scanners though GATE simulations.

Acknowledgements
Authors wish to thank Dr. Nektarios Kalyvas for his contribution in noise assessment.

5. References

[1] Riddell C, Carson R, Carrasquillo J, Libutti S, Danforth D, Whatley M and Bacharach S. 2001 *J. Nucl. Med.* **42** 1316.
[2] Liow J, Strother S. 1993 *Phys. Med. Biol.* **38** 55.
[3] Jaskowiak C, Bianco J, Perlman S and Fine J. 2005 *J. Nucl. Med.* **46** 424.
[4] Soret M, Bacharach S, and Buvat I. 2007 *J Nucl Med* **48** 932.
[5] Rapisarda E, Bettinardi V, Thielemans K and Gilardi M. 2010 *Phys. Med. Biol.* **55** 4131.
[6] Fin L, Bailly P, Daouk J and Meyer M. 2009 *Med. Phys.* **36** 3072.
[7] Martini N, Koukou V, Kalyvas N, Sotiropoulou P, Michail C, Valais I, Bakas A, Kandarakis I, Nikiforidis G and Fountos G. 2015 *J. Phys. Conf. Ser.* **574** 01207.
[8] Karpetas G, Michail C, Fountos G, Kandarakis I and Panayiotakis G 2014 *Nucl. Med. Com.* **35(8)** 967.
[9] Fountos G P, Michail C M, Zanglis A, Samartzis A, Martini N, Koukou V, Kalatzis I and Kandarakis I S 2012 *Med. Phys.* **39(3)** 1561.
[10] Fountos G, Zanglis A, Michail C, Kalatzis I, Cavouras D, Samartzis A, Kounadi E, Valsamaki P, Gerali S, Nikiforidis G and Kandarakis I, IFMBE Proceedings 25/II, pp. 802-805, 2009.
[11] Karpetas G E, Michail C M, Fountos G P, Valsamaki P N, Kandarakis I S and Panayiotakis G S 2013 *Hell J. Nucl. Med.* **16(2)** 111.
[12] Karpetas G E, Michail C M, Fountos G P, Kalyvas N I, Valais I G, Kandarakis I S, Panayiotakis G S 2014 *J. Phys.: Conf. Ser.* **490** 012139.
[13] OpenGATE Collaboration, Users Guide V6.1: http://www.opengatecollaboration.org.
[14] Thielemans K, Tsoumpas C, Mustafovic S, Beisel T, Aguiar P, Dikaios N and Jacobson M W 2012 *Phys. Med. Biol.* **57** 867.
[15] Valais I, Michail C, David S, Liaparinos P, Fountos G, Paschalis T, Kandarakis I, Panayiotakis G 2010 *IEEE Trans. Nucl. Sci.* **57(1)** 3.
[16] Valais I, Michail C, David S, Costaridou L, Nomicos C D, Panayiotakis G S, Kandarakis I 2008 *Physica Medica* **24** 122.
[17] Valais I G, Michail C M, David S L, Konstantinidis A, Cavouras D A, Kandarakis I S and Panayiotakis G S 2008 *IEEE Trans. Nucl. Sci.* **55(2)** 785.
[18] Michail C M, Fountos G P, Liaparinos P F, Kalyvas N E, Valais I, Kandarakis I S and Panayiotakis G S 2010 *Med. Phys.* **37(7)** 3694.
[19] Michail C, Kalyvas N, Valais I, David S, Seferis I, Toutountzis A, Karabotsos A, Liaparinos P, Fountos G and Kandarakis I 2013 *J. Lumin.* **144** 45.
[20] Boone J M 2001 *Med. Phys.* **28** 356.

Pediatric personalized CT-dosimetry Monte Carlo simulations, using computational phantoms

P Papadimitroulas[1], GC Kagadis[1], A Ploussi[2], S Kordolaimi[2], D Papamichail[3], E Karavasilis[4], V Syrgiamiotis[5], G Loudos[6]

[1] Department of Medical Physics, Medical School, University of Patras, GR 26504, Greece
[2] Department of Radiology, Medical School, National and Kapodistrian University of Athens, GR 12462, Greece
[3] Department of Radiology, Medical School, National and Kapodistrian University of Athens, GR 11528, Greece
[4] Radiology Research Unit, Medical Imaging Department, National and Kapodistrian University of Athens, Athens 11528, Greece
[5] MRI-CT Medical Imaging Department General Childrens' Hospital of Athens Ag. Sophia, Greece
[6] Department of Biomedical Engineering, Technological Educational Institute of Athens, GR 12210, Greece

E-mail: panospap@upatras.gr

Abstract. The last 40 years Monte Carlo (MC) simulations serve as a "gold standard" tool for a wide range of applications in the field of medical physics and tend to be essential in daily clinical practice. Regarding diagnostic imaging applications, such as computed tomography (CT), the assessment of deposited energy is of high interest, so as to better analyze the risks and the benefits of the procedure. The last few years a big effort is done towards personalized dosimetry, especially in pediatric applications. In the present study the GATE toolkit was used and computational pediatric phantoms have been modeled for the assessment of CT examinations dosimetry. The pediatric models used come from the XCAT and IT'IS series. The X-ray spectrum of a Brightspeed CT scanner was simulated and validated with experimental data. Specifically, a DCT-10 ionization chamber was irradiated twice using 120 kVp with 100 mAs and 200 mAs, for 1 sec in 1 central axial slice (thickness = 10mm). The absorbed dose was measured in air resulting in differences lower than 4% between the experimental and simulated data. The simulations were acquired using $\sim 10^{10}$ number of primaries in order to achieve low statistical uncertainties. Dose maps were also saved for quantification of the absorbed dose in several children critical organs during CT acquisition.

1. Introduction

Patient specific dosimetry is of high interest in pediatric applications where exposure to ionizing radiation is widely debated. The development of accurate anatomical virtual anthropomorphic models has reached a mature state for clinical exploitation. X-rays used in computed tomography (CT) for diagnosis can damage DNA directly or indirectly [1]. Although the radiation exposure per scan is decreased, the contribution of CT exams to collective dose is increasing rapidly due to the continuous rising in the frequency of CT examinations [2]. Pediatric patients are more susceptible to radiation-induced risks than are adult patients owing to their more rapidly growing tissues, their wider and increased cellular distribution of skeletal active marrow, and their greater post-exposure life expectancy. Recent studies conducted on a large pediatric population concluded that an increase in cancer incidence could be attributed to CT ionizing radiation, although the radiation doses are relatively small [3]. For that reason, the accurate calculation of pediatric radiation doses is vital in order to optimize CT protocols and to better understand the potential radiation risks of CT examinations. CT radiation doses are currently estimated by the volume computed tomography dose index ($CTDI_{vol}$) and the dose length product (DLP) dose indices.

Monte Carlo (MC) simulations is a widely recognized and well validated technique to study the particle interactions within the matter in nuclear medicine, radiology and radiation therapy. A plethora of MC codes have been developed especially since the early 90s. All the developed codes are mainly divided in two main categories, the “generic” and the “dedicated” codes. EGS, MCNP and GEANT4 are three of the most widely used and validated general purpose codes, while there are several MC tools for specific imaging or therapy simulations such as SIMSET, PET-SORTEO, TOPAS, GAMOS. GATE is a dedicated MC toolkit based on the Geant4 generic code [4]. Initially, it was designed for imaging single photon emission tomography (SPECT), and positron emission tomography (PET) applications while more recently it has been extended in Radiotherapy (RT) and CT imaging. We decided to use GATE in the present study, because it offers additional precision considering physics modeling and it has been extensively validated both on simple and realistic geometries in both dosimetry and imaging applications [5].

The use of anthropomorphic computational phantoms is rapidly gaining interest in medical sciences for the optimization of the acquisition and the processing protocols used in clinical practice. CT scans have a relatively high radiation exposure with associated long-term cancer risk, so computational anthropomorphic models provide an accurate and effective way to test and optimize the maximization of image quality and the minimization of absorbed dose per organ.

2. Materials & Methods

In the present study, the GATE open-source MC toolkit (v6.2) was used for modeling a 16 multidetector CT scanner Brightspeed, GE Healthcare. A DCT-10 ionization chamber was modeled in GATE in order to validate the simulated X-ray source. Furthermore, clinical pediatric CT data were processed (segmented) to be imported in GATE and the XCAT [6] and IT’IS [7] anthropomorphic phantoms were used as reference models for the dosimetric simulations.

In order to have accurate and realistic simulations the need of high number of simulated particles (a clinical study is in the order of $\sim 10^{13}$ particles) as well as fine resolution on the voxelized model ≤ 2.0 mm is apparent. The main limitation for acquiring such simulations is the computational cost. The final simulations were executed in the GateLab grid using 500 CPUs per simulation [8]. The execution time was almost 1 day per simulation (with $\sim 10^{10}$ simulated particles). The output of the simulations was 3D dose-maps, where the absorbed dose was stored in each voxel of the phantom. The dose-map loaded in MatLab as 3D matrix, and the dose per organ was calculated, according to the initial phantom.

2.1. Modelling the X-ray source

The initial goal was to set up and standardize the CT scanner model in GATE platform. There are three basic requirements in order to set up a whole simulation study for dosimetric applications. The exact source definition (energy, particles, angle, rotation, etc.) is necessary for the particle generation, and then the phantom description should be done taking into account its geometric characteristics and the material composition for particle tracking within it. Finally, the output file, i.e. “Dose Actor”, should be defined in order to store the deposited energy in each voxel within the phantom.

The source energy spectrum was calculated using the commercial software SpekCalc [9]. The calculated beam spectrum was inserted in the GATE platform as a general particle source (GPS), handled by Geant4. Specifically it was defined as a gamma-particle point source with an isotropic angular distribution equal to:

$$89.4^{\circ} \leq \varphi \leq 90.6^{\circ} \text{ and } 152.5^{\circ} \leq \theta \leq 207.5^{\circ},$$

where φ and θ stand for the spherical angle coordinates.

The energy photon spectrum was defined using a manual histogram of the energies (in MeV) and their intensity value in each respective histogram point (energy). In order to model the rotational movement of the CT X-ray beam, an extra file was prepared in the GATE simulator. In order to achieve an almost continuous movement of the source the 360° were divided in 800 steps of 0.45° angle rotation in each step (1.25 msec per step, for a full rotation of 1 sec). Defining more parameters in this file (Time, RotationAngle, RotationAxis and Translation in mm), a helical movement of the source was realistically modeled. In order to be as more realistic as possible, no variance reduction techniques (VRTs) were used in the modeling of the physical processes, nor cuts on the simulated physical volumes. All the appropriate physical processes that are taking place were modeled using the standard model of Geant4.

2.2. Methodology of absorbed dose in pediatric phantoms

In the current study 2 computational pediatric phantoms were modeled in GATE; a 5 years old girl from the IT'IS foundation and a 5 years old girl from the XCAT pediatric population. Both models were used as a reference for the quantification of the difference absorbed dose per organ between these different anatomical models (same age and gender). Clinical CT data were used to create computational models of individualized patients in order to move to personalized dosimetry. Specifically, 40 patients were used, 20 of which underwent head CT scans, while the rest underwent abdomen-pelvis CT examinations. After a semi-automated segmentation method the clinical data were converted to head and abdomen-pelvis phantoms for the simulator.

Our goal was to create a simulated pediatric population based on clinical data, which will be irradiated in GATE (helical irradiation) and the absorbed doses per organ (in mGy) in several groups of children will be given for reference to the scientific community.

3. Results

3.1. Verification of CT-source

The calculated energy spectrum compared to the simulated one is presented in figure 1.

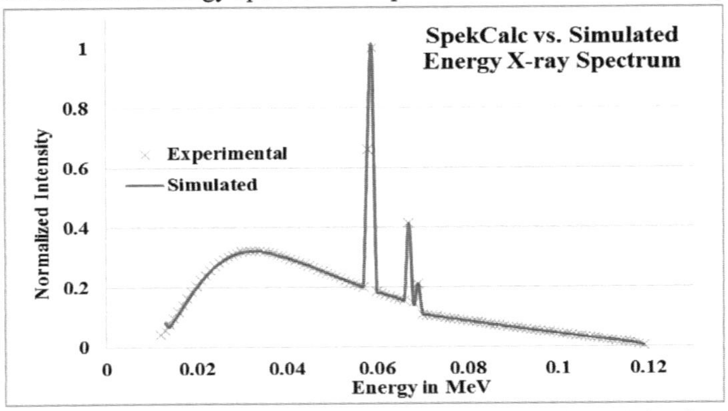

Figure 1. Comparison between calculated (SpekCalc) and simulated (GATE) spectrum.

Simulations were applied to quantify the absorbed dose a) in air, using a DCT-10 ionization chamber, and b) in PMMA, using a CTDI body phantom. Both in experiment and simulation irradiation was done in axial scans and in the center of the phantom. The results are shown in Table 1.

Table 1. Comparison between measured and simulated absorbed dose in CDTI

Phantom	*kVp*	*mAs*	*Experimental abs. Dose (mGy)*	*Simulated abs. Dose (mG)*	*Difference (%)*
Air	120	200	109.30	113.01	3.28
	120	100	54.50	56.51	3.56
PMMA	120	200	12.10	12.90	6.20
	120	350	22.90	24.10	5.02

3.2. Individualized pediatric dosimetry per organ

The methodology on extracting absorbed doses per organ was standardized. The output of GATE simulations is a 3D "Dose map" where the deposited energy per voxel (in MeV) is stored. Having the anatomical map of the model (attenuation map), the borders of the organs can be accurately defined in the dose output as shown in Figure 2.

Journal of Physics: Conference Series **637** (2015) 012020 doi:10.1088/1742-6596/637/1/012020

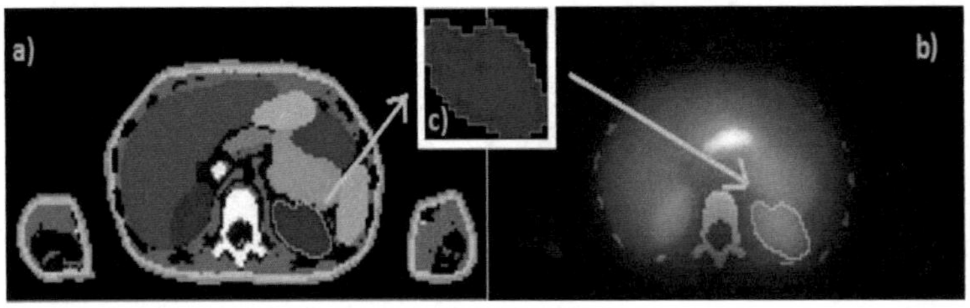

Figure 2. a) Phantom paediatric model with selected ROI, b) 3D simulated dose map, c) extracted organ to serve as ground truth for the boarders of the dose map.

All the simulations run with ~10^10 primaries as they resulted in statistical uncertainties lower than 8% per organ of interest.
The 2 computational models (5 years old XCAT and IT'IS) were irradiated in head, resulting in absorbed dose differences lower than 7% in brain.

4. Conclusion & Discussion

This is an ongoing study for optimization of pediatric dosimetry based on personalized clinical CT data. The verification of the CT scanner was completed, with high accuracy comparing clinical and simulated results, both in air and PMMA. The highest recorded difference was 6.20% in PMMA while the lower 3.28% in air. Initial dosimetric simulations were applied using standard pediatric phantoms, while a pediatric population was created based on clinical data. Our next step is to irradiate several children phantoms (head and abdomen-pelvis) in order to extract the absorbed doses in the most critical organs, taking into account the differentiation of the anatomy of each patient.

Acknowledgments

This research has been cofinanced by the European Union (European Social Fund) and Greek national resources under the framework of "Archimedes III: Funding of Research Groups in TEI of Athens" project of the "Education and Lifelong Learning" Operational Programme, and the John S. Latsis Public Benefit Foundation.

References

[1] Brenner DJ, Hall EJ 2007 Computed tomography-an increasing source of radiation exposure *N Engl J Med* **357** 2277-84
[2] Mettler FA Jr, Huda W, Yoshizumi TT, Mahesh M, 2008 Effective doses in radiology and diagnostic nuclear medicine: a catalog. *Radiology* **248** 254-63
[3] Vassileva J *et al* 2012 IAEA survey of pediatric CT practice in 40 countries in Asia, Europe, Latin America, and Africa: Part 1, frequency and appropriateness. *AJR Am J Roentgenol.* **198** 1021-31
[4] Jan S *et al* 2011 GATE V6: a major enhancement of the GATE simulation platform enabling modelling of CT and radiotherapy *Phys Med Biol* **56** 881-901
[5] Sarrut D *et al* 2014 A review of the use and potential of the GATE Monte Carlo simulation code for radiation therapy and dosimetry applications *Medical physics* **41** 064301
[6] Norris H, Zhang Y *et al* 2014 A set of 4D pediatric XCAT reference phantoms for multimodality research *Med Phys* **41** 033701
[7] Christ A *et al* 2010 The Virtual Family-development of surface-based anatomical models of two adults and two children for dosimetric simulations *Phys Med Biol* **55** N23-38
[8] Camarasu-Pop S, Glatard T, Da Silva RF, Gueth P, Sarrut D, Benoit- Cattin H, 2013 Monte Carlo simulation on heterogeneous distributed systems: A computing framework with parallel merging and checkpointing strategies *Fut Gen Comp Sys* **29** 728-38
[9] Poludniowski G, Landry G, DeBlois F, Evans PM, Verhaegen F, 2009 SpekCalc: a program to calculate photon spectra from tungsten anode x-ray tubes *Phys Med Biol* **54** N433-8

Journal of Physics: Conference Series **637** (2015) 012021 doi:10.1088/1742-6596/637/1/012021

A Low Cost Sensor Controller for Health Monitoring

M Birbas[1,2], **N Petrellis**[3], **F Gioulekas**[4]
[1]Analogies S.A., Patras Science Park, Patra-Rio, Greece
[2]Electrical and Computer Engineering Dept., University of Patras, Greece
[3]Computer Science and Engineering Dept., TEI of Thessaly, Larissa, Greece
[4]Sub-directorate of Informatics, University General Hospital of Larissa
Larissa, Greece

E-mail: npetrellis@teilar.gr

Abstract. Aging population can benefit from health care systems that allow their health and daily life to be monitored by expert medical staff. Blood pressure, temperature measurements or more advanced tests like Electrocardiograms (ECG) can be ordered through such a healthcare system while urgent situations can be detected and alleviated on time. The results of these tests can be stored with security in a remote cloud or database. Such systems are often used to monitor non-life threatening patient health problems and their advantage in lowering the cost of the healthcare services is obvious. A low cost commercial medical sensor kit has been used in the present work, trying to improve the accuracy and stability of the sensor measurements, the power consumption, etc. This Sensor Controller communicates with a Gateway installed in the patient's residence and a tablet or smart phone used for giving instructions to the patient through a comprehensive user interface. A flexible communication protocol has been defined supporting any short or long term sensor sampling scenario. The experimental results show that it is possible to achieve low power consumption by applying apropriate sleep intervals to the Sensor Controller and by deactivating periodically some of its functionality.

1. Introduction

A monitoring system has been proposed in [1][2] based on wireless sensors and smart phones to monitor the wellbeing of high risk cardiac patients. The work proposed in [3] introduces a perception layer, where a processing device incorporates wireless protocols to connect the patient sensors to the monitoring infrastructure and collect the corresponding data. These sensors produce raw values of data which are relayed to a central transceiver unit worn by the patient where they are converted into meaningful metadata. This work is focused on the protocol implementation rather than low-complexity and low-consumption sensors and integration platform. In [4] the available sensor technology and data analysis techniques are analysed and reviewed, towards the implementation of wearable systems for patient monitoring. Manufactured microelectromechanical systems (MEMS) are integrated with microprocessors and radio communication circuits. A new low-power ECG acquisition chip presented in [4] consumes only 680 nA when all features are active. It also provides competitive performance, such as input SNR greater than 70 dB and a Common Mode Rejection Ratio (CMRR) higher than 90 dB. In [5] an experimental Body Sensor Network platform measures important health indications and maps the health status of the individual carrying the platform into a set of predefined classes. The system presented in [5] is based on the Cooking Hacks e-Health development system used in this work. The work presented in [6] is based on a gateway that performs digital signal filtering and streaming to Web clients through a gateway. The gateway communicates with the microcontroller Raspberry PI in a wireless manner in order to collect the data from body sensors using another microcontroller board, the Arduino UNO.

We have also used as a sensor microcontroller, the low-cost eHealth kit of Libelium-Cooking Hacks for the implementation of an efficient e-health platform. Unlike other existing works we have managed to improve its key features by applying appropriate algorithms and software/hardware techniques. A low cost flexible e-health sensor network/platform has been developed that can support with increased reliability as many medical cases as possible. Low cost but with accuracy close to that offered by certified medical sensors is achieved. Low power consumption techniques have been employed by exploiting the idle intervals between measurements and transforming them to appropriate sleep intervals for the system and by deactivating part of the microcontroller functionality in a periodic manner in order to extend the battery life and the overall platform portability. A flexible control and communication protocol (between the Sensor Controller and the Gateway) has been developed to schedule the appropriate intervals that the various sensors should be sampled according to any user specification.

2. System Description and Sensor Controller

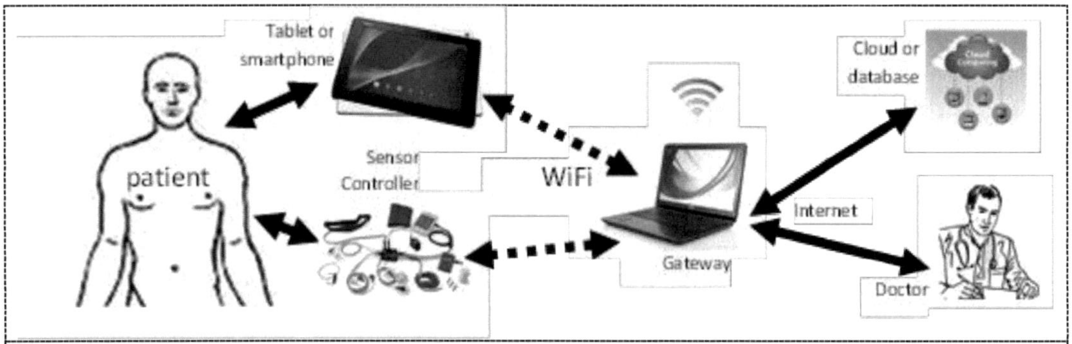

Figure 1. The framework of the health monitoring system.

Fig. 1 show the Sensor Controller developed in this work in the health monitoring framework. Instructions are given to the patient through a smart phone or a tablet with a comfortable user interface. The medical sensors that are used by the patient for simple tests (body temperature, respiratory measurements, skin conductance, etc) or more advanced ones (Electrocardiograms-ECG, Electromyograms-EMG) are connected to the Sensor Controller. The Sensor Controller communicates with a Gateway in the patient's residence.

The communication protocol between the Gateway and the Sensor Controller should ensure the privacy and security of the exchanged data. The communication protocol is responsible for transferring instructions from the Gateway to the Sensor Controller concerning the sensors that should be used and their sampling frequency. Different sensors may require different sampling profiles that should be supported in order to achieve reusability for the system. Analogue sensors may benefit from some kind of filtering (e.g., averaging) in order to achieve more accurate measurements.

The low power dissipation of the mobile Sensor Controller regardless of the microcontroller used is another target. Its portability demands power supply from a light battery that should last sufficiently long before being recharged. If the microcontroller board and the sensors cannot be modified, the only parameters that can be used as power optimization targets are the microcontroller operation and the wireless communication protocol. The microcontroller power dissipation can be significantly reduced if appropriate sleep intervals are defined between scheduled actions. However, there are certain limitations of the commercial microcontrollers that prevent an optimal low power operation.

The Gateway is providing an intermediate node between the patient, the doctor/hospital and the tablet/smartphone forwarding instructions and results between the Sensor Controller and final repository (cloud, hospital data base, etc). Some alarms and trivial decisions can be directly taken locally by the Gateway. The important medical decisions are taken of course by the doctor that is responsible for the specific patient supervision based on the collected data.
The sensors used in health monitoring systems can have analogue interface. Smart sensors can also be used with digital interface. The analogue sensors can be connected to Analogue Digital Converter (ADC) input channels. External ADCs can be used if the ADCs of the microcontroller cannot meet the sensor precision specifications. External analogue circuitry (e.g., voltage dividers, Wheatstone bridges, or linear low noise amplifiers) may be necessary for the adaptation of the output range of the sensor to the input range of the ADC.
If the digital sensors are autonomous microcontroller systems their measurement results can be simply read by the Sensor Controller or they can communicate wirelessly, directly to the Gateway. Smart sensors with digital interface can be connected to serial buses like Serial Peripheral Interface (SPI), I2C or Universal Synchronous-Asynchronous Receiver Transmitter (USART/UART). Nevertheless, the boards that are used for the (usually wireless) communication between the Sensor Controller and the Gateway may also need to be controlled by one of these interfaces and thus, conflicts may occur. Additional hardware switching developed in the context of the present work allows multiple sensors to be multiplexed on the same pins. An increased SRAM and Flash memory size may be also necessary to store firmware with high complexity and the sensor output values that may be subject to averaging / filtering or encryption procedures.
The system developed by the authors is based on the low cost Cooking Hacks eHealth kit that supports a number of sensors: temperature, skin conductivity, position (accelerometer), breathing airflow, ECG and EMG (EMG cannot operate concurrently with ECG), blood pressure (connected to UART), pulse and oxygen in blood, and glucometer (connected to UART, cannot operate at the same time with blood pressure sensor). The eHealth kit platform can be directly plugged onto an Arduino development board or to a Raspberry PI through an Arduino bridge where the Raspberry digital pins are rearranged in order to provide Arduino-compatible headers along with additional ADC inputs.
Raspberry PI is used as a Sensor Controller instead of Arduino due to its higher processing power (it is based on a 32-bit ARM core) and its increased RAM/ROM size compared to the 8-bit AVR microcontroller core of Arduino. The advanced GUI offered by the Raspbian OS makes possible the use of the Sensor Controller in stand-alone mode too. WiFi interface was selected as the communication method between the Sensor Controller and the Gateway instead of Bluetooth due to its higher reliability and the incompatibilities discovered between the Bluetooth card used on the Sensor Controller side and the smart phones tested.

3. Measurement Accuracy, Power Dissipation and Communication Protocol issues

A sensor calibration scheme proposed by the sensor manufacturer may not be applicable in real time and the measurements may vary depending on environmental conditions. If an accurate sensor calibration is not possible all the time, the health monitoring system may not achieve a medical certification for commercial use. Except from hardware improvements that include signal adaptation circuits (voltage level shifters, dividers, amplifiers) that are immune to temperature variations, the authors have applied a simple averaging scheme in order to stabilize the indications retrieved by analogue sensors like temperature, skin conductance, breathing airflow. The algorithm applied uses a moving window of e.g. 5 samples in order to estimate an initial average A_i. Then the initial samples that differ more than a threshold T from A_i are excluded from the final average A_f estimation. Using this simple technique the Normalized Mean Square Error (NMSE) between the initial values and the averaged ones were compared using a medically certified sensor as a reference. The NMSE error for the temperature sensor was improved by 5, while the corresponding error for the airflow and the skin conductance sensor was improved by 3.56 and 9.1 times respectively.

The overall power consumption of the Sensor Controller depends on the microcontroller software tasks, the power of the connected sensors, the wireless communication board, and the consumption of other interfaces. One way to reduce the power consumption is to deactivate interfaces not used (e.g., PL/HDMI) outputs or to periodically power down shields that are not operating all the time like the WiFi network interface. The Roving Networks RN-171 WiFi interface used in our case, allows a sleep mode where only 4uA are consumed in this mode instead of the approximately200mA (depending on the transmission power) consumed in normal operation. This sleep mode is entered automatically after a time interval set by a specific counter and wakes after an interval set by a second counter. The full Sensor Controller system dissipation may be as high as 1A if all the modules operate concurrently and a battery charge is needed every hour if for example a 1Ah battery is used.

The communication protocol between the Sensor Controller and the Gateway is defined so that low power consumption is achieved and any kind of sensor sampling profile is supported. More specifically, a command file is sent by the gateway with the definition of the sampling frequency of each sensor. The Sensor Controller samples the various sensors until the one with the slowest sampling rate is read once. Then the data values retrieved are sent back to the Gateway and a new command file is received. If the same sampling scheme has to be repeated, the Sensor Controller simply reads again the same command file. The intervals between consecutive sensor sampling are implemented either as short or long sleeps. During long sleeps, the Sensor Controller is fully turned off by an additional Arduino microcontroller packed with the Raspberry PI since the latter does not support hibernation. Short sleeps are implemented as intervals where the Wifi and other unnecessary peripherals are turned off. In demonstrative medical scenarios used (e.g., a daily 4min checkup where body position sensor is initially sampled for 2min to confirm a correct rest position, followed by a 2min ECG test), the battery charging interval was extended from 1h up to 32 days.

4. Conclusions

The accuracy of medical sensor measurements performed in health monitoring systems was discussed in this paper. A 3-10 times improvement in the indications of analogue sensors was achieved using a low complexity averaging scheme. The selection of an appropriate communication protocol between the sensor controller and the gateway allowed the support of any sensor sampling profile and the implementation of the sleep intervals defined in this protocol extended the battery life from 1h to several days.

5. References

[1] Gay V and Leijdekkers P 2007 A Health Monitoring System Using Smart Phones and Wearable Sensors *International Journal of ARM* **8(2)**

[2] Chan V, Ray P and Parameswaran N 2008 Mobile e-Health monitoring: an agent-based approach *IET Communications*, 223-230

[3] Mukherjee S, Dolui K, Datta S-K (2014) Patient Health Management System using e-Health Monitoring Architecture *Proceedings of the 2014 IEEE IACC (Gurgaon, 21-22 Feb. 2014)* 400 – 405

[4] Patel S, Park H, Bonato P, Chan L and Rodgers M (2012) A review of wearable sensors and systems with application in rehabilitation *Journal of NeuroEngineering and Rehabilitation* 9-21

[5] Khelil A, Shaikh F-K, Sheikh A-A, Felemban E, Bojan H (2014) DigiAID: A Wearable Health Platform for Automated Self-tagging in Emergency Cases. *Proc. of the 4th International Conference on Wireless Mobile Communication and Healthcare (MobiHealth), (Athens, Greece 2-3 Nov 2014)*

[6] Granados J, et al. (2014) Towards energy-efficient HealthCare: An Internet-of-Things architecture using intelligent gateways. *Proc. Of the 4th Int. Conf. on Wireless Mobile Communication and Healthcare (Mobihealth), (Athens, Greece 2-3 Nov 2014)*

Acknowledgments

This work was performed as part of ELTAB project (465435).

Automated DICOM metadata and volumetric anatomical information extraction for radiation dosimetry

D Papamichail[1], A Ploussi[2], S Kordolaimi[2], E Karavasilis[3], P Papadimitroulas[4], V Syrgiamiotis[5], E Efstathopoulos[2]

[1] 1st Department of Radiology, Medical School, National and Kapodistrian University of Athens, Vas. Sofias 76, Athens 11528, Greece

[2] 2nd Department of Radiology, Medical School, National and Kapodistrian University of Athens, Rimini 1, Athens 12462, Greece

[3] Radiology Research Unit, Medical Imaging Department, National and Kapodistrian University of Athens, Athens 11528, Greece

[4] Department of Medical Physics, Medical School, University of Patras, 15310, Rion, Greece

[5] MRI-CT Medical Imaging Department General Childrens' Hospital of Athens Ag. Sophia

dimitris.papamihail@gmail.com

Abstract. *P*atient-specific dosimetry calculations based on simulation techniques have as a prerequisite the modeling of the modality system and the creation of voxelized phantoms. This procedure requires the knowledge of scanning parameters and patients' information included in a DICOM file as well as image segmentation. However, the extraction of this information is complicated and time-consuming. The objective of this study was to develop a simple graphical user interface (GUI) to (i) automatically extract metadata from every slice image of a DICOM file in a single query and (ii) interactively specify the regions of interest (ROI) without explicit access to the radiology information system. The user-friendly application developed in Matlab environment. The user can select a series of DICOM files and manage their text and graphical data. The metadata are automatically formatted and presented to the user as a Microsoft Excel file. The volumetric maps are formed by interactively specifying the ROIs and by assigning a specific value in every ROI. The result is stored in DICOM format, for data and trend analysis. The developed GUI is easy, fast and and constitutes a very useful tool for individualized dosimetry. One of the future goals is to incorporate a remote access to a PACS server functionality.

1. Introduction

Patient-specific dosimetry gains considerable importance in medical field due to the increase use of ionizing radiation for diagnostic and therapeutic purposes and the radiation-associated risk cancer [1]. Nowadays, Monte Carlo (MC) simulations serve as a "gold standard" tool for patient-specific dosimetry.

In order to perform MC dose calculations, the first step is the modeling of the modality system with the exact definition of the ionizing radiation source (spectra, filtration, geometry) [2]. This step requires knowledge of the modality type and the acquisition parameters such as scan mode, pitch, rotation time and slice thickness. All necessary data for MC simulations are included in a DICOM (Digital Imaging and Communications in Medicine) file. Metadata in DICOM files contain information regarding patients' demographic characteristics, protocol settings, imaging parameters and a whole host of additional information essential for individualized dosimetry.

The second requirement for patient-specific dosimetry using MC simulation techniques is the creation of a voxelized anthropomorphic phantom [2]. Voxel-based phantoms are derived from patients' anatomical images. Each image dataset is segmented in order to specify the regions of interest (ROIs) in every single slice.

Although several DICOM extraction metadata tools [3,4] and contour delineation methods [5,6] have been proposed in the literature, few of them are freely available and can meet the aforementioned requirements.

The aim of this study was to develop a simple graphical user interface (GUI) to automatically extract and store information from DICOM metadata in a single query, as well as interactively delineating the regions of interest on anatomical images without explicit access to the radiology information system.

2. Materials and Methods

The software tool was developed on MATLAB environment (version 7.14, MathWorks, Inc.) [7]. The structure of the GUI has been designed to extract the sought-after information in a user-friendly and time saving functionality (Figure 1).

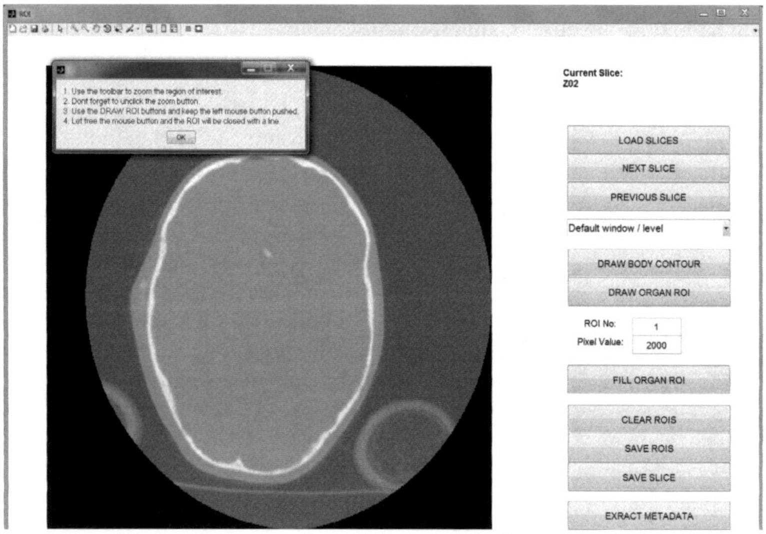

Figure 1. A snapshot of the graphical user interface.

Interactivity is the essence of the designed GUI [8]. In order to establish an interactive environment, two key components, the image display window and a series of buttons, form the GUI. Each button calls the corresponding computational module, developed in house. Images in DICOM format are processed as two-dimensional arrays (matrices). Each element of the matrix corresponds to a single pixel in the corresponding slice. All DICOM files should be retrieved from the PACS to a folder on a local hard-drive before the initialization of the program.

3. Results

3.1. Extracting metadata from multiple DICOM images

Starting the application, a dialog box enables the user to select multiple DICOM files from a folder. A computational module accesses and extracts automatically both metadata and image matrix of each file. By pushing the proper button, the application implements an algorithm gathering patient-, dose- and scan-related data from all selected files. The resulting information is automatically formatted and presented to the end user as a Microsoft Excel file for post-processing (Figure 2).

	A	B	C	D	E	F	G	H	I
1	Image	PatientAge	PatientWeight	SliceLocati	SliceThickı	SpacingBetweenSlice	KVP	XrayTubeCurrent	Exposure
2	Z01	010Y	22	6,455	5	20,705	120	254	63
3	Z02	010Y	22	11,631	5	20,705	120	254	63
4	Z03	010Y	22	16,808	5	20,705	120	254	63
5	Z04	010Y	22	21,984	5	20,705	120	254	63
6	Z05	010Y	22	27,16	5	20,705	120	230	57

Figure 2. A sample of the output file containing the metadata.

The data from the selected DICOM files are consolidated in the same worksheet for convenience. The accuracy of the results was tested by manual verification of contents seen on the DICOM metadata.

3.2. Specifying dosimetric regions of interest

In order to calculate radiation dose in Monte Carlo (MC) simulation, all regions of interest (ROIs) should be initially defined in each slice and subsequently be used as input for the simulations. Drawing contours in each slice is time consuming and requires a skilled user. A function to facilitate this process has been integrated in the developed application. The ROIs are defined interactively slice by slice, on the same image display window (Figure 3). The histogram of each image is automatically normalized, so that it is presented to the user in full-contrast range. If the boundaries of interest are not discrete, a desired range of intensity values (window level) can be specified, so that intensity is remapped to optimise the display of specific tissues.

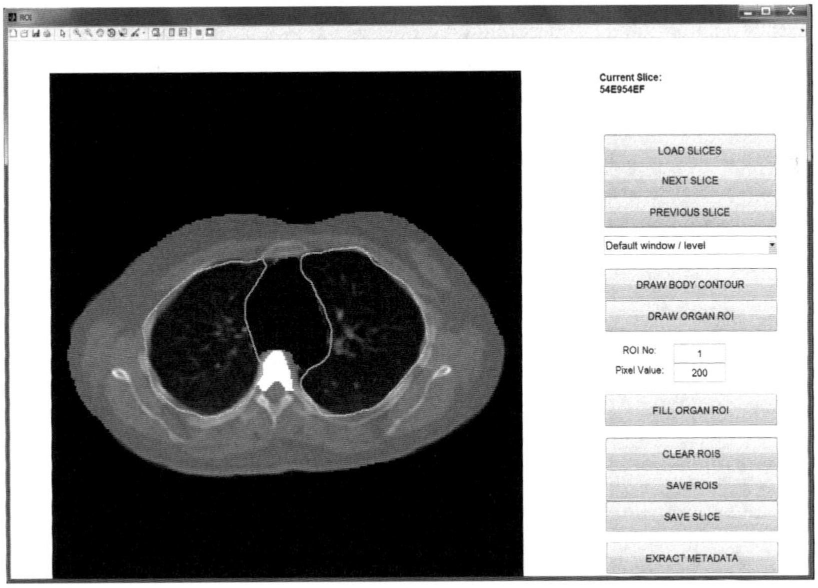

Figure 3. ROIs contouring in a CT slice of a PET/CT exam.

The user can specify several ROI boundaries in each slice by using an interactive pointer on the image display window. A specific value can be assigned in every ROI, so that all ROIs of the same organ can be easily matched at the end of the process. Each ROI can be deleted and redefined in any slice. ROIs of the same organ can be filled with a unique intensity number, so that the MC software can recognize each one as different tissue-equivalent material. Figure 4 depicts the flowchart of the algorithm developed.

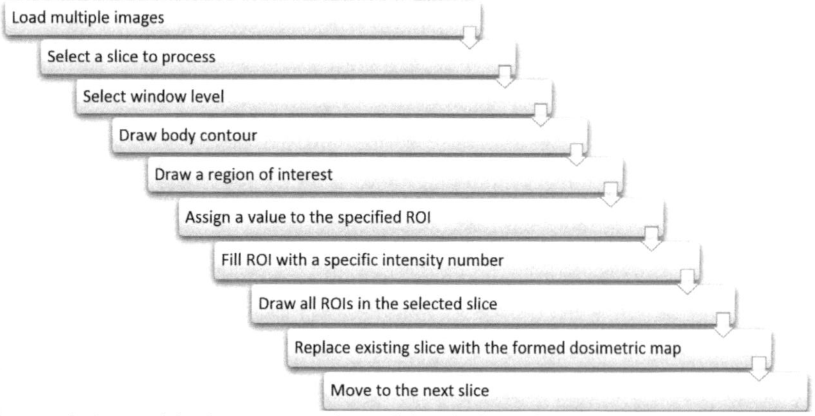

Figure 4. A graphical representation of the dosimetric map formation flowchart.

The interactive platform was successfully implemented in various computed tomography (CT) datasets like head, chest and abdomen-pelvis scans as well as for several modalities including angiography systems, γ-camera and positron emission tomography (PET).

4. Discussion

The proposed tool is designed to facilitate patient-specific dosimetry using MC simulations techniques. The application allows automated DICOM metadata extraction and manual contouring of ROIs. The performance of the application was evaluated by medical personnel for Radiology and Nuclear Medicine imaging modalities. The overall conclusion was that the software tool is effective, simple, fast and easy to use in clinical practice. The tool can be easily adapted for other modalities and examinations.

The use of the web-based platform is a step towards automatization of the pre-processing part of the data inserted in MC simulation codes saving time and effort in the whole process. The future goal of this study is the free disposal of the platform for use from a central server to any operating system.

Acknowledgement

The project was funded by the John S. Latsis Public Benefit Foundation. The sole responsibility for its content lies with its author.

References

[1] Mettler F, Huda W, Yoshizumi T and Mahesh M 2008 *Radiology* **248** 254–63
[2] De Marco J, Cagnon C, Cody D, Stevens D, McCollough C, O' Daniel J, McNitt-Gray M 2005 *Phys. Med. Biol.* **50** 3989–4004
[3] Li X, Zhang D, Liu B 2011 *AJR* **196** 781-83
[4] Dave J, Gingold E 2013 *AJR* **200** 142-45
[5] Zhou W, Xie Y 2014 *J Appl Clin Med Phys.* **15** 4499
[6] De Bruijne M, Van Ginneken B, Viergever M and Niessen W 2004 *Med Image Anal.* **8** 127–38
[7] Sobie EA 2011 *Sci Signal* **4** tr7
[8] Marchand P, Holland T 2003 *Graphics and GUI's with MATLAB* (Florida CRC Press)

Artificial Intelligence Methods Applied to Parameter Detection of Atrial Fibrillation

D Arotaritei and C Rotariu
Department of Biomedical Sciences, "Grigore T. Popa" University of Medicine and Pharmacy, Iasi, Romania

E-mail: cristian.rotariu@umfiasi.ro

Abstract. In this paper we present a novel method to develop an atrial fibrillation (AF) based on statistical descriptors and hybrid neuro-fuzzy and crisp system. The inference of system produce rules of type if-then-else that care extracted to construct a binary decision system: normal of atrial fibrillation. We use TPR (Turning Point Ratio), SE (Shannon Entropy) and RMSSD (Root Mean Square of Successive Differences) along with a new descriptor, Teager-Kaiser energy, in order to improve the accuracy of detection. The descriptors are calculated over a sliding window that produce very large number of vectors (massive dataset) used by classifier. The length of window is a crisp descriptor meanwhile the rest of descriptors are interval-valued type. The parameters of hybrid system are adapted using Genetic Algorithm (GA) algorithm with fitness single objective target: highest values for sensibility and sensitivity. The rules are extracted and they are part of the decision system. The proposed method was tested using the Physionet MIT-BIH Atrial Fibrillation Database and the experimental results revealed a good accuracy of AF detection in terms of sensitivity and specificity (above 90%).

1. Introduction

There are various methods proposed to detect atrial fibrillation (AF). The linear/nonlinear methods using Heart Rate Variability (HRV), Poincaré map, Spectral analysis, fractals and wavelets are the most used in publications.

A knowledge-based method is proposed in [1] using arrhythmic (ARH) episode detection for six rhythm types. The algorithm implies many rules and thresholds and a finite automaton is used for detection and classification. The algorithm reaches 98% accuracy for ARH classification and 94% accuracy for ARH classification and detection using MIT-BIH arrhythmia database. A methodology based on generalized discriminant analysis and for reduction of number of features is proposed in [2]. Artificial intelligence offers a variety of methods, tools and algorithms than can overcome some of the drawbacks of the classic methods. Among these methods, neural networks (NN) and evolutionary algorithms (especially genetic algorithms – GA) are the most popular. Some of the methods take the advantage of intelligent reasoning as in fuzzy systems and some combinations NN-fuzzy system proved to successful solution for classification problems. A feed forward NN was used in [3] as classifier for AF in three classes: sinus rhythm, supraventricular extra systoles and/or ventricular extra systoles. Fourier transform was used to observe the changes in QRS rhythm and NN was used to classify the rhythm in three classes with sensitivity and specificity > 98% [4]. Short-time generalized dimensions for cardiac rhythm characterization was proposed in [5].

In this paper we propose an evolutionary for parameter optimization using genetic algorithms (GA) for a situation inspired by [6].

2. Data Pre-processing

The most common linear methods used in HRV are based on RR intervals. Other combination of methods as TPR (Turning Point Ratio) and Shannon Entropy along with RMSSD have been proposed in [6]. We will use the definition of TPR, SE and RMSSD as they have been proposed in [6] along with a new descriptor, Teager-Kaiser (TK) energy in order to improve the accuracy of detection. TK energy has been used in very few papers related to HRV analysis [7]. TK is a nonlinear energy and it value increase if tone of signal increase at the same amplitude. The higher is frequency, the higher is the value of TK operator. We propose to use the formula given in [7], with regularization depending on length of window. We denote by x a sequence of n values.

$$TK\{x(k)\} = x^2(k) - x(k-1)\cdot x(k+1) \tag{1}$$

The average energy over a window of length L is defined by:

$$ATK_L = \left(\sum_i TK_i\right)/n^2 \tag{2}$$

where L is the window length, and n is the number of values in L window. The bigger is number of i's the higher is the sum ATK_L. The signals are first filtered from ectopic beats, reducing the source of false detection of AF.

The tradeoff between benefits and disadvantages in our case showed that is better to remove them. There are some proposal to filter ectopic beats and outliers ([6], [8], and [9]). We have been chosen the method from [6] but for treatment of outliers we proposed to replace the outlier the average value of left and right valid RR instead of remove them as in [6].

After filtration, all the RR segments are not linked together as in [5], but the files that contains the RR segments are individual records for the same data test. The set of parameters are focused on length of window L because an optimal window can produce a relevant value of randomness of time series that can be interpreted as AF [10]. A set P of k parameters over a window L can be defined as $P = [L, p_1, p_2, ..., p_k]$. Optimization of some parameters can be conflictual that is if optimization of objective using one parameter can produce a decrease of optimized value for other parameters. A tradeoff using optimized curve of values can be used to choose a correct tradeoff (e.g. Pareto front).

3. Optimization using genetic algorithms (GA)

The selected parameters are those from [6], L – length of window, Pth_1 – threshold from where the segment is considered to be AF (number of AF segments/number of total segments), RMSt – threshold of Root Mean Square of Successive Differences, TPRt threshold of Turning Point Ratio, SEt - threshold of Shannon Entropy and TKt - threshold of Teager-Kaiser energy. We must remark that usual method used in [6] is to split each interval in a number of slots and use all the combination to choose an optimal one that is a large number of repeated calculi.

The first question is if an individual HRV descriptor is enough for a good detection of AF, that is the sensitivity Se and specificity Sp are both enough close to 1.0 for a correct classification.

The main classification problem is that there is overlapping of N (Normal) values over AF (Atrial Fibrillation) values. A selection of type below is hard to be successfully defined. In these conditions, taking into account more descriptors can be a solution: it is possible that for a window of given length, an overlapping value of one descriptor that conducts to an imprecise decision to be solved using another descriptor for that decision is precise. This study will be developed in the further research.

The parameters used by the developed classifier are given by: $P = [L, P_{thresh}, mRMSSD_{thresh}, TPR_{thresh}, SE_{thresh}, ATK_{thresh}]$, where $mRMSSD = RMSSD/n$, a normalized RMSSD value.

We propose to use GA [11] for optimization in order to make both o more efficient and more compact procedure, easier to use for more than 6 variables. The operations are the known ones: selection using tournament method, crossover (in two points) and fitness calculation ([11]-[12]). The initialization of population with individuals is made using random generator. The last term, ATK is appending to improve the performance of decision system [6].

IF (mRMSSD > $mRMSSD_{thresh}$) AND (TPR in confidence interval TPR_{thresh}) AND (SE > SE_{thresh})
AND (ATK > ATK_{thresh}) THEN (segment$\in$AF, atrial fibrillation)
ELSE (segment$\in$N, normal).

The rule from above decision system is similar to [5] except the last decision ATK > ATKthresh.

4. Extraction of rules used in neuro-fuzzy inference system

Neuro-fuzzy systems are known to be universal approximators [13]. We propose to use ANFIS [14] as classifier for two classes of parameters: the class that identify the AN and the class for normal stage. The neuro-fuzzy inference system is used after that to extract fuzzy rules of type if-then-else and the membership functions are chosen to be triangular. In our approach, the output is a crisp value that are interpreted as discriminator between normal and AF segment.

There are many papers that propose to extract rules from structures that are proposed by artificial intelligence approach [15]. Decision trees are also used in medical decision system and in medical decision support system (DSS). Construction of decision tree can be based on crisp of fuzzy approach.

Automatic rules extraction from ANFIS fuzzy inference system is proposed to be developed in this paper. The output y is set to < 0.5 if segment of RR is normal and output y is set to ≥ 0.5 if segment of RR is associated with AF. In training stage, the values of thresholds for classification are set to 0.4 and 0.6 respectively.

5. Experimental results

The MIT-BIH Atrial Fibrillation database was used in our experiments. We select 9 files due to clarity of annotation of AF/N segments and the fact that record are complete with known start and no unread blocks. In the first approach we used the decision rules from [5], but the thresholds are identified using a systematic algorithm, heuristic optimization using GA. GA proved to be a good choice for optimization problem, single objective or multi-objective. The parameters are set to be limited by lower values [32, 0.1%, 0.01, 0.01%, 0.01, 0.01] and upper values [480, 99.99%, 0.999, 99.99%, 0.999, 0.999]. The chromosome coded for first value is mapping linearly into integer values between 32 and 480. The objective function is the maximum Se and Sp that is, these variables must have the closet values to 1.0. There are clearly to objective, possible contradictory, so the Pareto front offers to us a possibility to choose optimal values for both variables.

We start with a population of 50 individuals and after 40 iterations the solution is obtained. The multi-objective function is minimization of both functions 1-Se and 1-Sp. The results obtained are Se=90.17 and Sp = 90.12, are slightly less than those reported in [6], but there are using a less number of files. The values are acceptable for a good decision system. Moreover, the method can be extended to test more parameters avoiding possible explosive number of combinations as in [5].

Another test used the first five parameters without ATK. The results are slightly less performant, Se=90.01 and Sp = 88.76, and the proposed improvement by adding ATK proved that the new descriptor drive on to a better results. In this case, 21 windows of length 320 (the optimal one) were misclassified and as follow the performance is decreased.

6. Conclusions

We proposed to use GA to optimize values of parameters for detection of AF. A new parameter, Teager operator is introduced to optimize the detection. The selection of parameters is based on assumption that in AF, somewhat randomness characteristics are developed in RR records. We

proposed to use a systematic method to optimize the parameter thresholds for a decision system. Parameters can increase with minor modification of the algorithm that is, adding new descriptors supposed to be in two limits to detect AF, a simple new parameter is added in GA and the simulation software can run with minimal modifications.
The rules are however not optimized in the case of ANFIS inference that is the universe of discourse for each parameter is split in the same number of membership functions (e.g. 3). The algorithm for pruning rules and number of membership function will be developed in the further research.

Aclnowledgement
The work has been carried out within the program Joint Applied Research Projects, funded by the Romanian National Authority for Scientific Research (MEN – UEFISCDI), contract PN-II-PT-PCCA-2013-4-0761, no. 21/2014 (SIACT).

References
[1] M.G. Tsipouras, D.I. Fotiadis, and D. Sideris, An arrhythmia classification system based on the RR-interval signal, *Artificial Intelligence in Medicine*, vol. 33, pp. 237-250, 2005.
[2] B. M. Asl, S. K. Setarehdan, and M. Mohebbi, "Support vector machine-based arrhythmia classification using reduced features of heart rate variability signal, *Artificial Intelligence in Medicine*," vol. 44, pp. 51-54, 2008.
[3] T.F.Yang, B. Devine, and P.W. Macfarlane, "Artificial neural networks for the diagnosis of atrial fibrillation," *Med Biol Eng Comput.*, vol. 32, no. 6, pp. 615-619, 1994.
[4] K. Minami, H. Nakajima, and T. Toyoshima, Real-time discrimination of ventricular tachyarrhythmia with Fourier transform neural network, *IEEE Trans Biomed Eng*, vol. 46, no. 2, pp. 179-185,1999.
[5] Y. Wang, Y.S. Zhu, N.V.Thakor, and Y.H. Xu, A short-time multifractal approach for arrhythmia detection based on fuzzy neural network, *IEEE Trans Biomed Eng*, vol. 48, no. 9, pp.989-995, 2001.
[6] S. Dash, K.H. Chon, S. Lu, E.A. Raeder, Automatic Real Time Detection of Atrial Fibrillation, Annals of Biomedical Engineering, vol. 37, issue 9, pp. 1701-1709, 2009.
[7] M. D'Alessandro, et. all., "Epileptic seizure prediction using hybrid feature selection over multiple intracranial EEG electrode contacts: A report of four patients," *IEEE Transactions on Biomedical Engineering*, vol. 50, no. 5, pp. 603-615, 2003.
[8] R. Karlsson, R. Hörnsten, A. Rydberg, and U. Wiklund, Automatic filtering of outliers in RR intervals before analysis of heart rate variability in Holter recordings: a comparison with carefully edited data, *BioMedical Engineering OnLine*, pp. 1:12, 2012.
[9] N. Storck, M. Ericson, L. Lindblad, and M. Jensen-Urstad M, Automatic computerized analysis of heart rate variability with digital filtering of ectopic beats, *Clin. Physiol.*, vol. 21, no. 1, pp. 15-24, 2001.
[10] J.H. Zar, Serial Randomness of measurements: nonparametric testing, In: *Biostatisical Analysis*, Prentice-Hall, pp. 587-588, 1985.
[11] Z. Michalewicz, *Genetic Algorithms + Data Structures = Evolution Programs*, Springer, 1996.
[12] D. Goldberg, *Genetic Algorithms in Search, Optimization and Machine Learning*, Reading, MA: Addison-Wesley Professional, 1989.
[13] Castro, J.L., Delgado, M.: Fuzzy systems with defuzzification are Universal approximators. In: *IEEE Transactions on Systems, Man and Cybernetic*, vol. 26, no.1, pp. 149-152 (1996)
[14] J.-S. Roger Jang, ANFIS: Adaptive-network-based fuzzy inference system, *IEEE Transactions on System, Man, and Cybernetics*, No. 5, Vol. 23, PP. 665-685, May 1993.
[15] S. Kulluk, L. Özbakır, A. Baykasoglu, Fuzzy DIFACONN-miner: A novel approach for fuzzy rule extraction from neural networks, *Expert Systems with Applications* vol. 40 pp. 938–946, 2013.

Pediatric chest HRCT using the iDose[4] Hybrid Iterative Reconstruction Algorithm: Which iDose level to choose?

M Smarda[1], E Alexopoulou[1], A Mazioti[1], S Kordolaimi[1], A Ploussi[1], K Priftis[2] and E Efstathopoulos[1]

[1]2nd Department of Radiology, Medical School, National and Kapodistrian University of Athens, Rimini 1, Athens 12462, Greece

[2]Department of Pediatrics, University General Hospital ATTIKON, Chaidari, Athens, Greece

stathise@med.uoa.gr

Abstract. Purpose of the study is to determine the appropriate iterative reconstruction (IR) algorithm level that combines image quality and diagnostic confidence, for pediatric patients undergoing high-resolution computed tomography (HRCT). During the last 2 years, a total number of 20 children up to 10 years old with a clinical presentation of chronic bronchitis underwent HRCT in our department's 64-detector row CT scanner using the iDose IR algorithm, with almost similar image settings (80kVp, 40-50 mAs). CT images were reconstructed with all iDose levels (level 1 to 7) as well as with filtered-back projection (FBP) algorithm. Subjective image quality was evaluated by 2 experienced radiologists in terms of image noise, sharpness, contrast and diagnostic acceptability using a 5-point scale (1=excellent image, 5=non-acceptable image). Artifacts existance was also pointed out. All mean scores from both radiologists corresponded to satisfactory image quality (score ≤ 3), even with the FBP algorithm use. Almost excellent (score <2) overall image quality was achieved with iDose levels 5 to 7, but oversmoothing artifacts appearing with iDose levels 6 and 7 affected the diagnostic confidence. In conclusion, the use of iDose level 5 enables almost excellent image quality without considerable artifacts affecting the diagnosis. Further evaluation is needed in order to draw more precise conclusions.

1. Introduction

The large number of computed tomography (CT) examinations taking place in everyday clinical practice has increased the risk for radiation-induced carcinogenesis and has made CT radiation dose a nowadays top safety concern in health care. This concern is much stronger for children who are more radiosensitive and have longer life expectancy than adults [1].

The use of iterative reconstruction (IR) algorithms in CT examinations has been proved useful for reducing radiation dose intake, while preserving image quality [2, 3]. However, there still remains a question which IR algorithm level to choose in order to combine minimal image noise with maximum diagnostic acceptability, as higher IR levels have been proved to be connected with blotchy, pixelated or oversmoothing artifacts providing a 'non-natural' image [3, 4].

The purpose of this study is to determine the appropriate Hybrid iDose IR algorithm level for the chosen image settings (mAs, kVp) used in our radiology department in children often in need for repetitive chest high-resolution computed tomography (HRCT) examinations with the indication of chronic suppurative lung disease.

2. Materials and Methods

The present cohort study had the approval of our hospital ethics committee.

2.1 Patient Population

We retrospectively reviewed preschool and preadolescent school-aged pediatric patients, under the age of 10 years old, who underwent chest HRCT examinations after the installation of iDose[4] IR algorithm in our Radiology Department. All the patients suffered from chronic suppurative lung disease.

According to the review, a total number of 24 children with the indication of chronic suppurative bronchitis underwent the specific examination since 2012 using the iDose IR algorithm. For technical reasons and mainly because of poor cooperation and severe motion artifacts, 4 of them were excluded. Therefore, a total number of 20 pediatric patients finally participated in the study **(Table 1)**.

Table 1. Patients' characteristics. Data are presented as mean ± standard deviation and range in parentheses.

Characteristics	
Age (yr)	5.53 ± 1.35 (4-9.5)
No of patients	20
Male/Female	7/13
Weight (kg)	19.92 ± 3.06 (14.7-28)

2.2 Acquisition protocol

All the HRCT examinations were performed in a 64-slice multidetector CT (MDCT) scanner (Brilliance, Philips Healthcare, Cleveland, OH, USA), using 80kVp voltage and 40-50 mAs settings. In addition, all examinations were reconstructed using both FBP algorithm and all iDose levels (levels 1 to 7), in order to draw conclusions for the appropriate kVp, mAs settings and iDose level combination.

HRCT scanning protocol included non-contiguous axial slices (every 1cm) from lung apex to base for insiratory scans, and limited (3 to 4) spaced HRCT slices for expiratory scans, using 1.25mm slice thickness. The rest CT scanning parameters included supine patient positioning, 64 x 0.625mm detector configuration, 0.5sec gantry rotation time and a sharp filter (YC – Philips Healthcare, Cleveland, OH).

2.3 Subjective Image Quality

The subjective image quality of all the aforementioned examinations was evaluated independently by 2 experienced radiologists (A with 22 years of experience and B with 8 years of experience) for noise, sharpness, contrast and diagnostic acceptability using a descending five-point scale **(Table 2)**. According to this scale grade 1 corresponds to excellent image quality, grade 5 corresponds to non-acceptable image quality, whereas grade 3 constitutes the threshold for diagnostic acceptable image. The overall image quality was estimated as the average value of image noise, sharpness and contrast scores. The presence of artifacts (respiratory, metallic, oversmoothing, blotchy or pixelated appearance etc.) affecting the diagnosis was also pointed out.

Table 2. Subjective image quality evaluation

	Noise	**Sharpness**	**Contrast**	**Diagnostic acceptability**
1	No noise	Excellent sharpness of structures	Excellent contrast	Superior
2	Minimal noise	Good sharpness of structures	Very good contrast	Good
3	Average noise, not affecting the diagnosis	Acceptable sharpness	Acceptable contrast	Average
4	Above average noise, affecting the	Difficulty in distinguishing	Slight contrast enhancement	Suboptimal

	diagnosis	structures		
5	Unacceptable noise	Unacceptable	Very poor contrast	Unacceptable

2.4 Statistical analysis
Interobserver agreement between the two radiologists was evaluated using kappa (k) test with k< 0.20 signified poor agreement; k=0.20–0.40, fair agreement; k=0.41–0.60, moderate agreement; k=0.61–0.80, good agreement; and k=0.81–1.0, almost perfect agreement. The statistical test was performed using the Superior Performance Software System (SPSS, v 21.0, SPSS Inc., Chicago, IL, USA). Subjective image quality data were presented as mean scores from both radiologists.

3. Results
The interobserver agreement showed a good agreement between the 2 independently working radiologists (k>0.628). There also was an expected progressive improvement in image quality from iDose level 1 to iDose level 7 reconstruction algorithm for all the aforementioned parameters (noise, sharpness, contrast and diagnostic acceptability).

A remarkable notice was that all mean scores from both radiologists were from 1 to 3, even with the FBP algorithm use, which is considered an acceptable score as it corresponds to a satisfactory image quality, without important noise affecting the diagnosis. Almost excellent image quality (score <2) was obtained with iDose levels 5, 6 and 7.

The presence of artifacts affecting the diagnosis was also assessed, with the majority relating to respiration or motion artifacts and in one case artifacts from metallic implants. The existence of these artifacts in most of the cases affected the diagnostic acceptability, but not to such an extent as to render the examinations non-diagnostic (scores up to 3). Additionally, artifacts responsible for blotchy or pixelated appearance described in literature were not observed in this study, whereas oversmoothing artifacts coming from the use of high levels of iterative algorithms were only noticed with iDose levels 6 (6/20 of children) and 7 (8/20 of children). Despite the fact that in our study oversmoothing artifacts did not have a negative impact on the subjective evaluation of contrast or sharpness (scores almost equal to 1), whenever noticed an issue of imaging interpretation arose, because the final CT image outcome was far from our radiologists' everyday diagnostic experience and familiarization. In these cases, the radiologists who participated in our study remained slightly reserved in scoring. But since these artifacts were only observed in a small number of children, mean diagnostic acceptability evaluation of iDose levels 6 and 7 practically was not affected, as shown in table 3.

Table 3. Subjective image quality evaluation. Data are presented as mean scores from both radiologists-observers participating in the study.

	FBP	iDose1	iDose2	iDose3	iDose4	iDose5	iDose6	iDose7
Noise	2.9	2.8	2.5	2.3	1.9	1.8	1.4	1.2
Sharpness	2.8	2.7	2.6	2.2	1.9	1.6	1.4	1.1
Contrast	2.6	2.5	2.3	2.1	1.7	1.5	1.4	1.1
Diagnostic acceptability	2.5	2.4	2.4	2.2	2.1	1.9	1.6	1.4
Overall image quality	2.6	2.6	2.5	2.1	1.9	1.6	1.4	1.1

4. Discussion
As already mentioned, the aim of this study was to determine the appropriate iDose level providing excellent image quality for the chosen kVp and mAs settings in children's chest HRCT examinations. Two more studies for chest CT with a similar reasoning have been conducted so far, one for adults by Singh et al. [5], using 30%, 50%, and 70% of Adaptive Statistical Iterative Reconstruction technique (ASIR) blending, and one by B. Karmazyn et al. [6] for children's chest-abdomen with FBP and iDose

levels 2 to 6. The first one [5] revealed mild pixilated blotchy texture only with 70% blended ASIR images and the second one [6] showed that iDose levels 3 and 4 where optimal for most cases.

According to the subjective image quality evaluation of our study, excellent subjective image quality (score equal or almost equal to 1) was only achieved with the use of iDose level 7 and almost excellent image quality (score <2) was obtained with iDose levels 5 and 6. However, among the aforementioned levels, iDose level 5 was the only one not accompanied by oversmoothing artifacts. The existence of these artifacts, as already explained, affected our experienced pediatric radiologists' diagnostic confidence by leaving them with doubts concerning CT image reliability. This resulted in them preferring for diagnosis a lower iDose level with less good scores concerning noise, contrast or sharpness, over higher iDose levels with an admittedly excellent overall image quality (average of noise, contrast and sharpness scores).

An important observation as well, was that subjective overall image quality was always evaluated up to 3 (acceptable image quality), even with the use of the FBP algorithm. According to this observation, future investigation with further reduction of mAs settings could be achieved without affecting the diagnostic outcome.

Among the limitations of this particular study concerning chest HRCT with the use of iDose IR algorithm for chronic suppurative bronchitis was the small number of the participating pediatric patients. This suggests that so far the advantages of the IR technique are not sufficiently widespread in the medical world and the concern of pediatricians towards young patients' radiation exposure still remains. The limited number of patients participating in the study can be partially attributed as well to the fact that there were non-cooperative children during the CT scans. These children had to be excluded from the study due to significant motion artifacts, as they were considered to be affecting the results concerning diagnostic acceptability.

In conclusion, our study results imply that iDose level 5 could provide almost excellent image quality for our radiology department's chosen kVp and mAs settings, without any substantial artifacts affecting the diagnostic outcome. Further evaluation with even more patients and lower mAs settings could lead to safer conclusions for this 'sensitive' patient group in need for repetitive chest HRCT examinations.

Acknowledgement
iDose4 software and hardware were courteously provided by PHILIPS Healthcare within the auspices of PHILIPS CT Publication of the year award won by our department in 2009.

References
[1] Ogbole G I 2010 Radiation dose in pediatric computed tomography: risks and benefits *An Ibadan Post med.* **8** 118-26
[2] Hara A K et al. 2009 Iterative reconstruction technique for reducing body radiation dose at CT: feasibility study *Am J Roentgenol.* **193** 764-71
[3] Singh S et al. 2012 Radiation dose reduction with hybrid iterative reconstruction for pediatric CT *Radiol.* **263** 537-46
[4] Lee S H et al. 2012 Radiation dose reduction with the adaptive statistical iterative reconstruction (ASIR) technique for chest CT in children: an intra-individual comparison *Eur J Radiol.* **81** 938–43
[5] Singh S et al. 2011 Adaptive statistical iterative reconstruction technique for radiation dose reduction in chest CT: a pilot study *Radiol.* **259** 565-73
[6] Karmazyn B et al. 2014 Optimization of hybrid iterative reconstruction level in pediatric body CT *Am J Roentgenol.* **202** 426-31

X-ray dual energy spectral parameter optimization for bone Calcium/Phosphorus mass ratio estimation

P. I. Sotiropoulou[1], G. P. Fountos[2], N. D. Martini[1], V. N. Koukou[1], C. M. Michail[2], I. G. Valais[2], I. S. Kandarakis[2] and G. C. Nikiforidis[1]

[1] Department of Medical Physics, Faculty of Medicine, University of Patras, 265 00 Patras, Greece
[2] Radiation Physics, Materials Technology and Biomedical Imaging Laboratory, Department of Biomedical Engineering, Technological Educational Institute of Athens, Egaleo, 122 10 Athens, Greece

E-mail: gfoun@teiath.gr

Abstract. Calcium (Ca) and Phosphorus (P) bone mass ratio has been identified as an important, yet underutilized, risk factor in osteoporosis diagnosis. The purpose of this simulation study is to investigate the use of effective or mean mass attenuation coefficient in Ca/P mass ratio estimation with the use of a dual-energy method. The investigation was based on the minimization of the accuracy of Ca/P ratio, with respect to the Coefficient of Variation of the ratio. Different set-ups were examined, based on the K-edge filtering technique and single X-ray exposure. The modified X-ray output was attenuated by various Ca/P mass ratios resulting in nine calibration points, while keeping constant the total bone thickness. The simulated data were obtained considering a photon counting energy discriminating detector. The standard deviation of the residuals was used to compare and evaluate the accuracy between the different dual energy set-ups. The optimum mass attenuation coefficient for the Ca/P mass ratio estimation was the effective coefficient in all the examined set-ups. The variation of the residuals between the different set-ups was not significant.

1. Introduction

Osteoporosis has been recognized as a worldwide disease that affects more than 75 million people in the United States, Europe and Japan [1,2]. Several methods [2,3] are available for measuring an individual's bone mineral status. Bone mineral density (BMD) is the gold standard measurement, determining the bone health, which primarily measures the quantity of bone in the skeleton. However, about one-half of fractures occur in women with a T-Score above the World Health Organization (WHO) diagnosis threshold of osteoporosis [1,4], suggesting that other factors, besides bone quantity, are involved in the appearance of fractures. The major and essential component of human bones is hydroxyapatite ($Ca_{10}(PO_4)_6(OH)_2$), a mineral form of calcium (Ca) and phosphorus (P) that gives bones their rigidity. Researchers [2,5,6] have determined alterations in skeletal Ca/P molar ratio. There are indications [5-10] that the alterations on Ca/P ratio have a negative effect on bone health.

In this study, an optimization in the Ca/P bone mass ratio estimation was held based on the dual energy method, previously described by in previous works of our group [7,8,11]. Simulated data were obtained using the K-edge filtering technique and single X-ray exposure. The modified X-ray output was attenuated by nine calibration points produced by various Ca/P mass ratios. The analysis presented here, investigates the use of effective or mean mass attenuation coefficient in Ca/P mass ratio estimation. The optimization was based on the minimization of the accuracy of Ca/P ratio, with respect to the coefficient of variation of the Ca/P ratio ($CV_{Ca/P}$). The standard deviation of the residuals was used to compare and evaluate the accuracy between the different dual energy set-ups.

2. Materials and Methods

2.1 Simulation input functions and parameters

The theory for the *in vivo* determination of Ca/P ratio, in the radius, was described previously [8,11]. The Ca/P mass ratio is obtained assuming that there is a three-component system: Ca, PO_4 and water (w). The photon beam is transmitted through an object of total thickness t, $t = t_{Ca} + t_{PO_4} + t_w$ where t_{Ca} is the thickness of calcium, t_{PO4} is the thickness of phosphate and t_w is the thickness of water. The total thickness was equal to 5cm, corresponding to a human forearm. Assuming that both surfaces of Ca and PO_4 were exposed to the same radiation beam and considering that the molecular weight ratio PO_4 / P is 3.0679, the Ca/P mass ratio [7,11] is given by:

$$\frac{Ca}{P} = \frac{t_{Ca}}{t_{PO_4}} \cdot 3.0679 \tag{1}$$

In a previous study [11], a wide range of filter materials and thicknesses, combined with different maximum tube voltages (kVps), were simulated producing the different set-ups. The set-ups were used for the dual energy spectra optimization, in the Ca/P mass ratio estimation. The simulation considers a photon counting energy dispersive detector that allows energy peak discrimination and counting.

2.2 Mass attenuation coefficients

In Table 1 the step thicknesses for Ca and PO_4 are shown, used to generate the detector data. The three pairs of Ca and PO_4 thicknesses correspond to molar ratios of 1.40, 1.58, and 1.67, respectively. The combined thicknesses results in 9 calibration points, where the total thickness of Ca, PO_4 and water is maintained constant.

Table 1. Ca and PO_4 calibration step thicknesses (in cm) used to generate simulated detector calibration data.

Material	Thickness (cm)		
Ca	0.0533	0.0671	0.0704
PO_4	0.1168	0.1017	0.0996

In the accuracy (α) determination [11] of the Ca/P mass ratio (Eq. 1), instead of the effective mass attenuation coefficient μ/ρ_{eff} that was used in the previous study [11,12], the mass attenuation coefficient, corresponding to the mean energy of the incident spectrum (mean mass attenuation coefficient, μ/ρ_{ME}), can be used to estimate the Ca/P mass ratio. The accuracy of the estimated Ca/P mass ratio values was calculated for both μ/ρ_{eff} and μ/ρ_{ME} values.

For every calibration point, the residuals (δ) were calculated as the difference between the true (tr) and the estimated values (est) according to following equation:

$$\delta = Ca/P_{tr} - Ca/P_{est} \tag{2}$$

The standard deviation of the residuals (δ_{rms}) was used to compare and evaluate the accuracy between the different dual energy set-ups. δ_{rms} values close to zero indicates that the set-up has a smaller random error component. Different set-ups were examined, based on the K-edge filtering technique and single X-ray exposure [13]. This study used three set-ups with optimum $CV_{Ca/P}$ described in a previous study [11]. These are La (0.15cm, 100 kVp) (set-up S1), Ce (0.13cm, 100 kVp) (set-up S2) and Sm (0.13cm, 100 kVp) (set-up S3).

3. Results and Discussion

The accuracy values were calculated using both the effective (μ/ρ_{eff}) and mean (μ/ρ_{ME}) mass attenuation coefficients. For the set-ups S1, S2 and S3, the accuracy ranged from 2.31 to 5.18%, using μ/ρ_{eff} for the calculations, and from 8.33 to 18.98% using μ/ρ_{ME} , respectively. Set-up S3 appeared to have the lower accuracy value (2.31%), as well as the optimum δ_{rms} (0.014) value. The results for set-ups S1, S2, S3 are shown in Table 2. The δ_{rms} values in three set-ups close to zero (Table 1) suggest that the errors from the Ca/P estimation will be sufficiently small in the calibration region. δ_{rms} values between the different set-ups were comparable. This implies that the accuracy values of each set-up are the main indicator for the optimum set-up and spectra optimization.

Table 2. δ_{rms} and accuracy range for the examined set-ups calculated with μ/ρ_{eff} and μ/ρ_{ME} .

	μ/ρ_{eff}		μ/ρ_{ME}	
	δ_{rms}	a% (range)	δ_{rms}	a% (range)
set-up S1	0.033	3.60-4.22	0.067	15.47-18.98
set-up S2	0.019	4.25-5.18	0.066	15.49-18.61
set-up S3	0.014	2.31-2.97	0.077	8.33-12.31

The use of μ/ρ_{eff} instead of μ/ρ_{ME} provides results with lower accuracy values. The accuracy values can be up to 6 times higher if calculated with μ/ρ_{ME} . In Table 3 the values of μ/ρ_{eff} and μ/ρ_{ME} are given for set-up S1. The μ/ρ_{ME} values are lower from the μ/ρ_{eff} values in all cases. These differences, in μ/ρ_{eff} and μ/ρ_{ME} , justify the accuracy differences in Ca/P estimation.

Table 3. Mass attenuation coefficients of Ca, PO_4, w for set-up S1.

	μ/ρ_{eff}	μ/ρ_{ME}
$Ca_{(Ehigh)}$	0.7589	0.6866
$Ca_{(Elow)}$	2.2670	2.2570
$PO_{4\,(Ehigh)}$	0.2574	0.2458
$PO_{4\,(Elow)}$	0.5082	0.5069
$w_{(Ehigh)}$	0.2105	0.2072
$w_{(Elow)}$	0.2897	0.2897

The data from radius are included in this study solely as a proof-of-concept demonstration. The accuracy and the precision of measurements from bone phantoms and specimens is the subject of ongoing work in our lab.

4. Conclusions

Analysis of the bone composition increases sensitivity and specificity in metabolic bone diseases diagnosis and their monitoring. In this study an optimization in the Ca/P bone mass ratio estimation was investigated using the either the effective or mean mass attenuation coefficients. The use of the μ/ρ_{eff} instead of μ/ρ_{ME} provides results with lower accuracy values.

Acknowledgements

This research has been co-funded by the European Union (European Social Fund) and Greek national resources under the framework of the "Archimedes III: Funding of Research Groups in TEI of Athens" project of the "Education & Lifelong Learning" Operational Program.

References

[1] Ruppel M, Miller L, Burr D 2008 *Osteoporos. Int.* **19** 1251
[2] Kyriazis V and Tzaphlidou M. 2004 *Scien. World J.* **4** 1027
[3] Travert C, Jolivet E, Brosses ES, Mitton D, Skalli W 2011 *Med Biol Eng Comput* **49** 1355
[4] Sornay-Rendu E, Munoz F, Garnero P, Duboeuf F, Delmas PD 2005 *J Bone Miner Res* **20** 1813
[5] Kourkoumelis N, Balatsoukas I, Tzaphlidou M 2012 *J Biol Phys* **38** 279
[6] Kounadi E, Fountos G, Tzaphlidou M 1998 *Connect. Tissue Res.* **37** 69
[7] Fountos G,Yasumura S, Glaros D 1997 *Med. Phys.* **24** 1303
[8] Fountos G, Tzaphlidou M, Kounadi E, Glaros D 1999 *Appl. Radiat. Isot.* **51** 273
[9] Neues F and Epple M 2008 *Chem. Rev.* **108** 4734
[10] Zoehrer R, Perilli E, Kuliwaba JS, Shapter NL, Fazzalari NL, Voelcker NH 2012 *Osteoporos Int* **23** 1297
[11] Sotiropoulou P, Fountos G, Martini N, Koukou V, Michail C, Kandarakis I and Nikiforidis G 2015 *Phys. Medica* **31** 307
[12] Sotiropoulou P, Fountos G, Martini N, Koukou V, C Michail C, Valais I, Kandarakis I and Nikiforidis G 2014 X-ray spectra for bone quality assessment using energy dispersive counting and imaging detectors with dual energy method XIII Mediterranean Conf. on Medical and Biological Engineering and Computing 2013 (Seville, Spain, 25-28 September) (IFMBE Proceedings Vol 41) ed LM Roa Romero (Springer Int.) p 463
[13] Martini N, Koukou V, Michail C, Sotiropoulou P, Kalyvas N, Kandarakis I, Nikiforidis G, Fountos G 2015 *J. of Spectr.* Article ID 563763, 8 pages

An evaluation of the clinical potential of tissue diffraction studies

R Speller[1], S Abuchi[1], Y Zheng[1], N Vassiljev[1], A Konstantinidis[2] and J Griffiths[1,3]
[1]Department of Medical Physics and Biomedical Engineering, UCL, London WC1E 6BT, UK
[2]Diagnostic Radiology and Radiation Protection, Christie Medical Physics and Engineering, The Christie NHS Foundation Trust, Manchester M20 4BX, UK
[3]Institute of Biomedical Engineering, UCL, London WC1E 6BT, UK

Email: r.speller@ucl.ac.uk

Abstract. Medical imaging is a long established part of patient management in the treatment of disease. However, in most cases it only provides anatomical detail and does not provide any form of tissue characterisation. This is particularly true for X-ray imaging. Recent studies on tissue diffraction have shown that true molecular signatures can be derived for different tissue types. Breast cancer samples and liver tissue have been studied. It has been shown that diffraction profiles can be traced away from the primary tumour in excised breast tissue samples and that potentially 3mm fat nodules in liver tissue can be identified in patients at acceptable doses.

1. Background

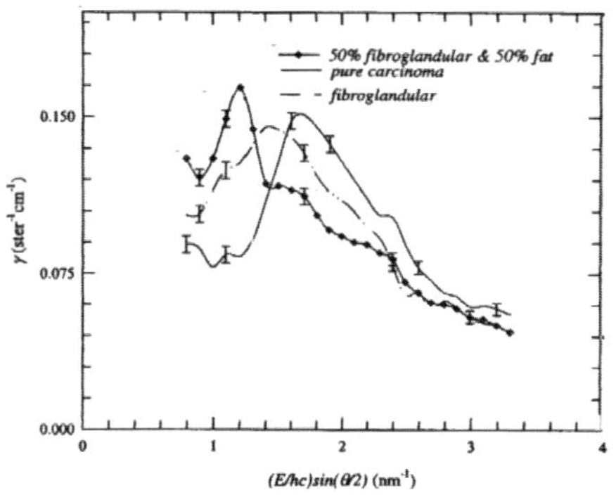

Figure 1: Diffraction signatures (indirect measurements of molecular form factors) for different breast tissues.

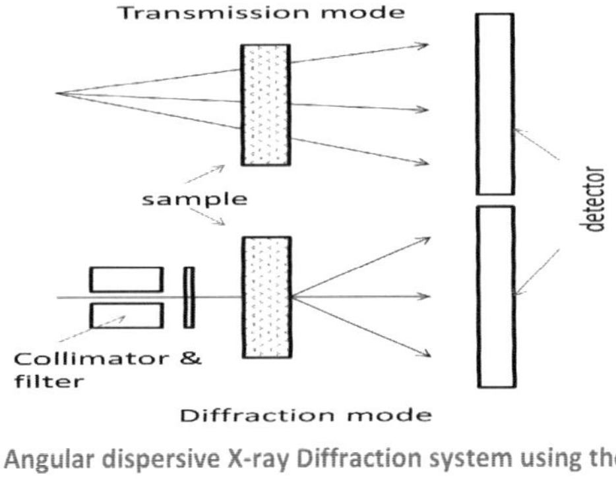

Figure 2: Angular dispersive X-ray Diffraction system using the DynAMITe detector. The sample – detector distance was 135mm and the X-ray beam reduced to ~2mm diameter.

One of the common medical diagnostic procedures is an X-ray examination. In most cases the image information is interpreted by considering subtle changes in X-ray transmission leading to information about the shapes and internal features in different organ structures. Considerable experience is required to convert this information into a diagnosis that can be used for patient management. One aspect is that the shape and appearance of a given structure is the only information available for classifying the tissue and with early stages of many diseases, changes in shape and appearance can be very subtle. This makes classifying a given tissue as diseased difficult.

Tissue diffraction using X-rays has, over the past 20 years, shown potential for classifying different disease states for a range of organs or tissues and hence could be used to quantify the status of disease. For example Figure 1 is for breast tissue (taken from Kidane *et al.* (1999)). In the current paper we look at two clinical problems and suggest how tissue diffraction measurements made with two different diffractometers could aid patient management - firstly breast cancer surgery and analysis of a wide local excision (WLE) and secondly assessment of fat nodules in liver tissue.

2. Breast tissue analysis

Early breast cancer is usually treated by breast conserving surgery, chemotherapy and radiotherapy. However, despite advances in modern personalised treatments the recurrence rates are generally ~20-40%[1]. The problem is thought to be due to inadequate removal of tissue during surgery. One possible approach to ensure complete removal whilst minimising the tissue removed is to study the diffraction profiles associated with suspicious regions in the WLE. To study this, several WLEs have been analysed with the diffraction system shown in Figure 2. The system consists of a tungsten target X-ray tube with a collimator system that can be introduced remotely to reduce the X-ray beam from a wide cone to a 2mm diameter pencil beam. Without the collimator, a conventional transmission image can be recorded on the detector system. When the collimator is introduced the pencil beam can be used to analyse small, selected regions of tissue where a diffraction pattern is recorded on the same detector used for transmission imaging. Generally two images are taken in the diffraction mode using two different filter materials, i.e. 0.1mm Samarium (Sm) and 0.2mm Tin (Sn). This combination of filters (i.e. balanced filtration) gives an effective narrow band of X-ray energies when the two images are subtracted. The detector system used for these measurements is called DynAMITe (Dynamic range Adjustable for Medical Imaging Technology) - see [2]. Typical results are shown in Figure 3. Figure 3(a) shows part of a high resolution transmission image of a breast tissue WLE. The primary tumour is in the upper part of the image. One of the infiltrating branches of the tumour has been subjected to diffraction analysis at a series of points (P1 to P7) indicated in the figure. At each of these points a diffraction signature has been measured and these are shown in Figure 3 (b). To obtain the 1D profiles shown in Figure 3(b) the 2D diffraction pattern recorded by the detector has been radially averaged. The radius of

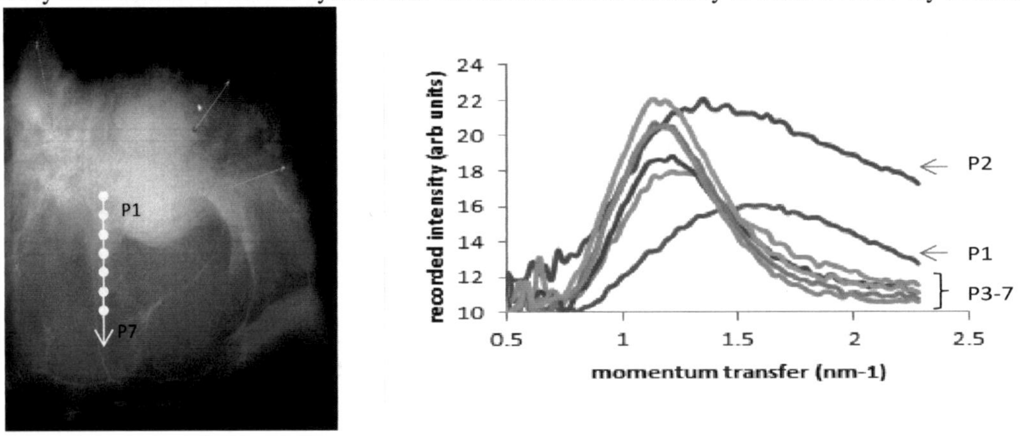

Figure 3: Diffraction analysis of a WLE tissue sample. (a) high resolution transmission image, (b) diffraction signatures.

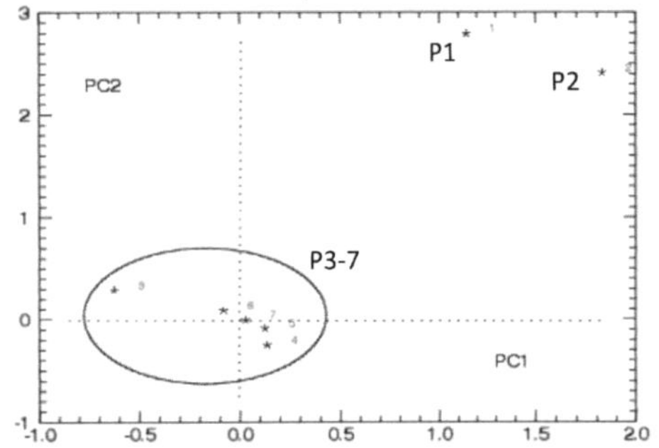

Figure 4: PC1 vs PC2 for the diffraction profiles shown in Figure 3.

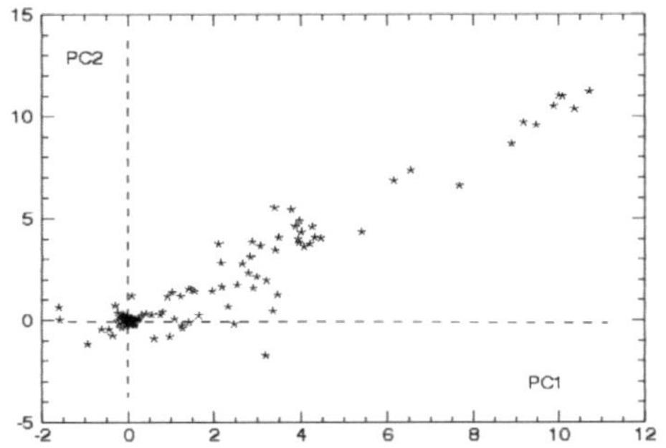

Figure 5: PC1 vs PC2 for diffraction profiles from 41 samples. These profiles were taken from the centre of the primary tumour moving towards the periphery in 1mm steps.

integration has been converted to momentum tranfer using the scattering angle and an estimate for the mean energy of the X-ray beam using the balanced filters described above. It can be seen that the diffraction signatures change as the point of measurement moves along the structure in the image. Comparison with Figure 1 suggests these changes in the profiles are related to the fraction of different tissues present at each point. Peaking at lower values of momentum transfer are the profiles measured some distance from the primary tumour whereas those from within the tumour have peaks close to $1.5nm^{-1}$. Principal component analysis (PCA) shows that more than 97% of variance is explained by PCAs 1 and 2. The scores plot given in Figure 4 shows clear separation between diffraction signatures (L3P3-L3P7) and L3P1 and L3P2. In total 41 samples have been looked at and results are shown in Figure 5. All breast tissue samples were taken from breast wide local excisions provided by the Tissue Bank at St. Bartholomew's Hospital, London, UK, under ethical approval from the local Research Ethics Committee (East London and The City REC Alpha). All patients had given informed consent for the use of redundant tissue from excision specimens for research purposes. Comparing Figures 4 and 5 suggests that tissue away from the primary tumour occupies a region close to the origin of the plot whereas tumour involvement moves towards the top right hand corner. Thus analysing a tissue sample removed during surgery to ensure the borders are clear of tumour would require the analysis to place the result close to the origin of Figure 5.

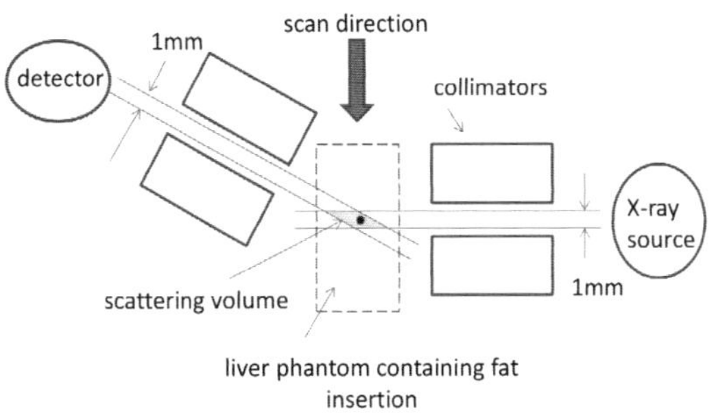

Figure 6: The diffraction system used for the analysis of liver samples. The scattering angle is 5^0

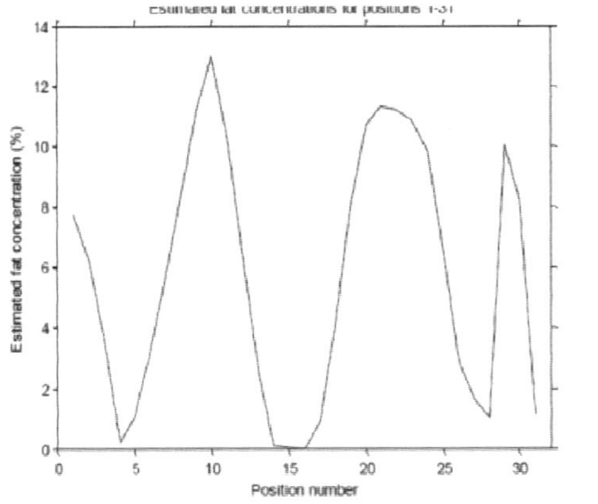

Figure 7:The predicted percentage of fat occupying the scattering volume as the sample is scanned across the X-ray beam. 31 steps separated by 1mm were used.

3. Fat content in liver tissue. Identifying normal tissue at the borders of a WLE is an investigation on an excised tissue sample, generally small in size. Quantifying the fat content of liver tissue would be a measurement made upon a patient. Hence, in this work, the analysis of liver tissue was approached in a different way. Small test phantoms were built and evaluated using a transmission energy dispersive X-ray diffraction system. The clinical potential was investigated by modelling based upon the results from the small-scale phantoms. The diffractometer uses a tungsten target X-ray source operated at 70kVp, a set of collimators that restrict scattering to 5^0 and an HpGe spectrometer. Figure 6 shows the equipment set-up. The two collimators allow phantoms up to 40mm thick to be examined. Frozen pig liver had different diameter cylinders of lard (4,5 and 6mm) inserted. Diffraction profiles were collected at 1mm intervals as the phantom was scanned a total of 31 different positions across the X-ray beam. The set of diffraction profiles were subjected to PCA analysis. The residuals, following subtraction of the mean of the diffraction profiles, generate curves that have the same shape but differ in their scales. This is typical behaviour of a two-component, additive system. Hence when PCA models the data, it estimates the shape of the common curve (loading vector) and the multiples of that curve present in each individual diffraction profile (the scores). These factors were then used to build a model to predict the percentage of fat present at each measurement point. The result of these estimations is shown in figure 7. Fat inclusions of 6, 5 and

4mm are shown as peaks at positions 10, 22 and 30 respectively. The predicted percentage of fat depends upon the diameter of the insert and the fraction of the scattering volume the insert occupies. The scattering angle was 5^0 leading to a total length for the scattering volume of ~23mm. As the inserts are significantly less than the scattering volume, a partial volume effect reduces the predicted occupancy to significantly less than 100%. To estimate whether or not these measurements could be made in a patient, the diffraction profiles measured with the phantoms were used as input to a model that allowed additional tissues to be added. The effect of these additional tissues would be to attenuate both the incident and scattered beams and to generate additional multiple scatter. The simple model ignored the multiple scatter but attempted to include the statistical quality of the data. Attenuating the beam would mean less photons would be recorded and hence the statistical quality of the data reduced unless the input flux, and hence dose, increased. Hence a range of patient sizes and a range of effective doses were simulated and the results subjected to the same analysis. The results are shown in Figure 8. Each curve is the result of the analysis described above divided by the result given in Figure 7. The top left hand corner of Figure 8 is for the smallest sized patient and the highest dose. As the resultant curve is a straight, horizontal line it means there no differences compared to the result in Figure 7, i.e., the small experimental phantom. Moving across the set of curves displayed in Figure 8 the patient becomes larger (going from child to large adult), moving down the set of curves the dose is reducing (going from an entrance surface dose of (3.2mGy to ~0.32mGy). The lower right hand corner (larger patients, lower doses) demonstrates large differences in detecting the fat inserts as their locations are incorrectly predicted. In the upper left section the curves only show small departures from predicting correctly. The results suggest that for conditions to the left of the dotted line in Figure 8 useful estimates of fat content could be made.

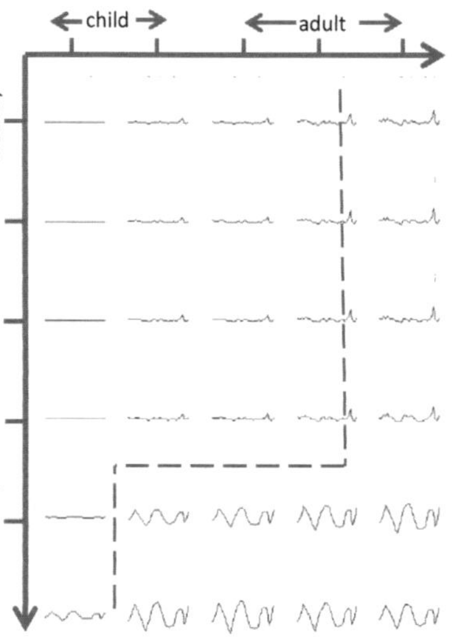

Figure 8: Modelled results for 5 different sized patients and 6 different dose levels using the experimental result in Figure 7 as input to the model. To highlight any changes from the detectability demonstrated in Figure 7 each result has been divided by the result in Figure 7.

4. Conclusions

Tissue diffraction measurements can provide useful clinical data. It has been shown that tissue removed during breast cancer management could be analysed in a way that could demonstrate that all the tumour has been removed. Our preliminary results show an ability to identify normal tissue from tumour in WLE. Analysing liver tissue to quantify the fat content has also been studied. The results, from a combination of experimental measurements and modelling, show that estimates at acceptable doses are only likely in the case of small adults or children.

Acknowledgements The authors would like to acknowledge the help of Prof. Tom Fearn and Dr Dan O'Flynn with respect to the PCA analysis of the liver data. NV is supported by the Rosetrees Foundation.

References

[1] McCahill L E, Single R M, Aiello Bowles E J, Feigelson H S, James T A, Barney T, Engel J M and Onitilo A A 2012 Variability in Re-excision Following Breast Conservation Surgery, *JAMA*. vol 307 pp 467-475.

[2] Kidane G, Speller R D, Royle G J and Hanby A M 1999 X-ray scatter signatures for normal and neoplastic breast tissues *Physics in Medicine and Biology* vol 44 pp 1791–1802

[3] Esposito M, Anaxagoras T, Konstantinidis A C, Zheng Y, Speller R D, Evans P M, Allinson N M and Wells K 2014 Performance of a novel wafer scale CMOS active pixel sensor for bio-medical imaging *Physics in Medicine and Biology* vol 59 pp 3533–355

Dynamic Infrared Thermography Study of Blood Flow Relative to Lower Limp Position

I. Stathopoulos[1], K. Skouroliakou[2], C. Michail[1] and I. Valais[1]

[1]Department of Biomedical Engineering, Technological Educational Institute of Athens, 12210 Egaleo Athens, Greece
[2]Department of Energy Technology Engineering, Technological Educational Institute of Athens, 12210 Egaleo Athens, Greece

E-mail: valais@teiath.gr

Abstract. Thermography is an established method for studying skin temperature distribution. Temperature distribution on body surface is influenced by a variety of physiological mechanisms and has been proven a reliable indicator of various physiological disorders. Blood flow is an important factor that influences body heat diffusion and skin temperature. In an attempt to validate and further elucidate thermal models characterizing the human skin, dynamic thermography of the lower limp in horizontal and vertical position was performed, using a FLIR T460 thermographic camera. Temporal variation of temperature was recorded on five distinct points of the limp. Specific points were initially cooled by the means of an ice cube and measurements of the skin temperature were obtained every 30 seconds as the skin temperature was locally reduced and afterwards restored at its initial value. The return to thermal balance followed roughly the same pattern for all points of measurement, although the heating rate was faster when the foot was in horizontal position. Thermal balance was achieved faster at the spots that were positioned on a vein passage. Our results confirm the influence of blood flow on the thermal regulation of the skin. Spots located over veins exhibit different thermal behaviour due to thermal convection through blood flow. Changing the position of the foot from vertical to horizontal, effectively affects blood perfusion as in the vertical position blood circulation is opposed by gravity.

Keywords: Dynamic Thermography, Thermographic Camera, Infrared, Thermograms

1. Introduction

The skin is a protective covering of the human body and a complex organ that contains several mechanisms to assist its good functioning. With its strength, elasticity, and cohesion of the whole structure, it protects the body against mechanical shocks and regulates temperature against the environmental temperature fluctuations [1]. Blood flow plays an important role in heat distribution. With modern experimental techniques heat distribution assessment became more precise. Infrared (IR) imaging or thermography is a method for acquiring infrared radiation associated with the spatial distribution of heat on the examined objects, and the variation of this distribution over time [2]. A typical thermographic camera detects radiation with a spectral range between 9-14 μm and relates intensity of radiation with temperature [1]. However the accuracy of thermography is limited due to the influence of factors other than temperature, such as position, vascularization, clothing, climate, etc. To better understand the thermomechanical behaviour of the skin and how it is influenced from its direct environment, we present an experimental study using dynamic infrared thermography [3]. This experimental study aims to quantitatively highlight effects of blood flow on heat diffusion. To this aim, infrared temperature measurements on the skin of a human leg were acquired to relate the tissue vascularity and leg position with leg temperature [1-4].

2. Materials And Methods

In an attempt to validate and further elucidate thermal models characterizing the human skin, dynamic thermography of the lower limp in the horizontal and vertical position was performed, using a FLIR T440 thermographic camera (FLIR Systems Inc., North Billerica, MA). This camera has a $25^0 \times 19^0$ field of view and an uncooled microbolometer focal plane array of 320×240 pixels. The spectral range of the camera is 7.5–13 μm. The thermal sensitivity (i.e. temperature resolution) is 0.045^0C at 30^0C ambient temperature. The absolute temperature measured depends on factors such as. The emissivity coefficient of the object's surface, ambient temperature and humidity. (Relevant parameters can be changed in the software after the images are recorded). The temperature accuracy is ±0,5 ^{0}C and ±2% if all the variables (emissivity, temperature and humidity) are correctly set. The values selected for the parameters mentioned above are presented in Table I. Relative temperature variations are used throughout this study, since minor temperature changes appear between the spots of each image, possibly due to human factor. Therefore, the inaccuracies in absolute temperature, due to dependence on emissivity or any other parameter settings are avoided [5].

Table 1. Selected values of adjustable imaging parameters.

Parameter	Value
Emissivity	1.00
Distance (m)	1.00
Temperature (^{0}C)	29.0
Relative Humidity (%)	50.0

Temporal variation of temperature was recorded on five distinct points of the limp. Specific points on the foot were initially cooled by means of an ice cube and measurements of the skin temperature were obtained every 30 seconds as the area temperature was locally reduced and afterwards restored at its initial value.

3. Results and Discussion

Figure 1a shows the average temporal temperature restoration with the foot in vertical and horizontal positions. Although average temperature restoration curves exhibit similar behavior, significant deviation can be observed at localized points (spots) especially at those over surface vessels (Figure 1b).

Figures 2(a) and 2(b) are infrared pictures of the temperature distribution of the feet in vertical position before (2a) and after (2b) the positioning of the ice cube at the right foot (Sp1) respectively.

doi:10.1088/1742-6596/637/1/012027

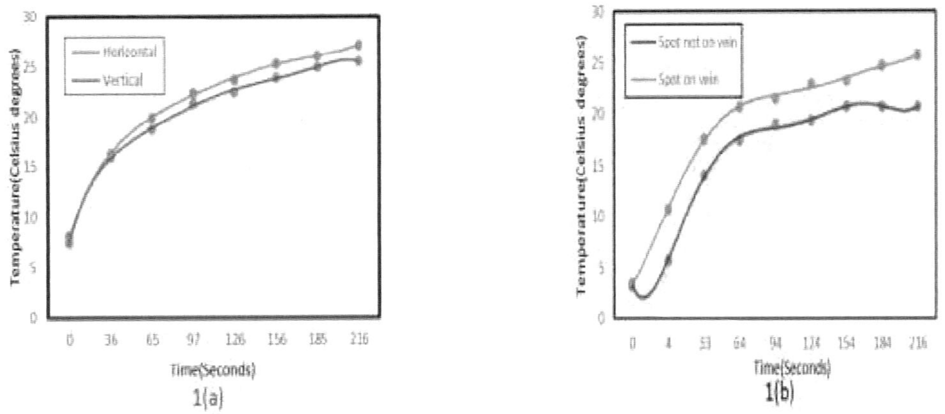

Figure 1: a) Temporal temperature restoration with the foot in vertical and horizontal position, b)Temporal temperature restoration for two spots (A and B), one of them located over a vein passage.

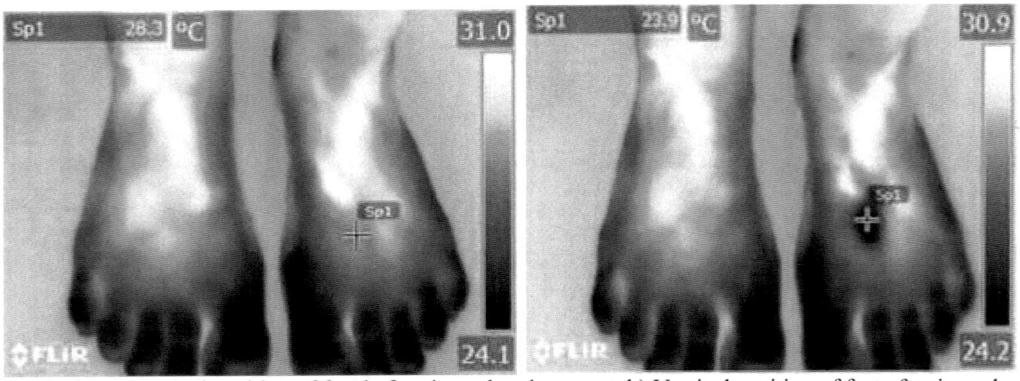

Figure 2. a) Vertical position of feet before ice cube placement, b) Vertical position of feet after ice cube placement

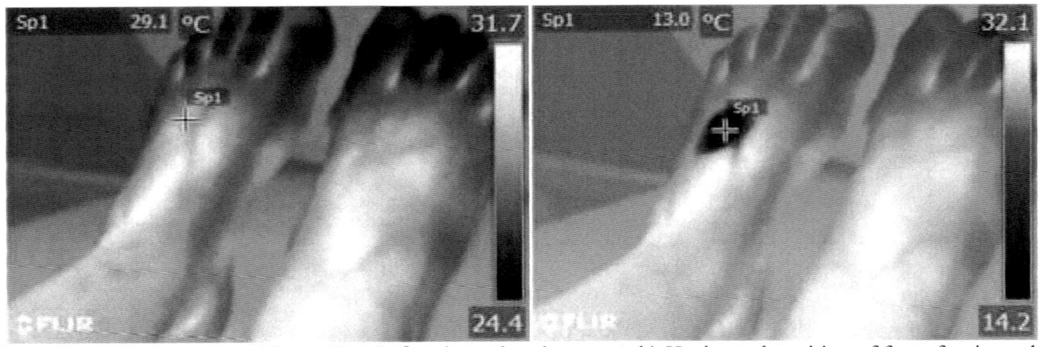

Figure 3. a) Horizontal position of feet before ice cube placement, b) Horizontal position of feet after ice cube placement

Figures 4(a) and 4(b) show the return to thermal balance after initial cooling of a foot area. All points of measurement followed roughly the same pattern, although the heating rate was faster when the foot was in horizontal position. Thermal balance was achieved faster at the spots that were positioned on a vein passage. The thermal response of spot which was under the cooled zone and on the passage of a vein, was like spots in cooled zone that was not on vein passage but with a higher initial temperature and a faster return to thermal balance. For the upward vertical position, observations are exactly the same as in the horizontal position. However, looking at spots in cooled area in the horizontal position, the return to thermal balance is faster. [1]

Differences in temperature variation due to different blood perfusion was also observed by Skouroliakou *et al*, 2014 in the anatomical area of the upper limb. Although thermographic absolute temperature measurement depends on many factors, relative temperature variation measured on the same spot of the same surface is reliable. As it can be seen from figures 4(a, b) gravity opposition to blood flow influences heat transfer mechanisms through blood flow. Thermal balance exhibit a faster trend at spots positioned over a vein passage. [5]

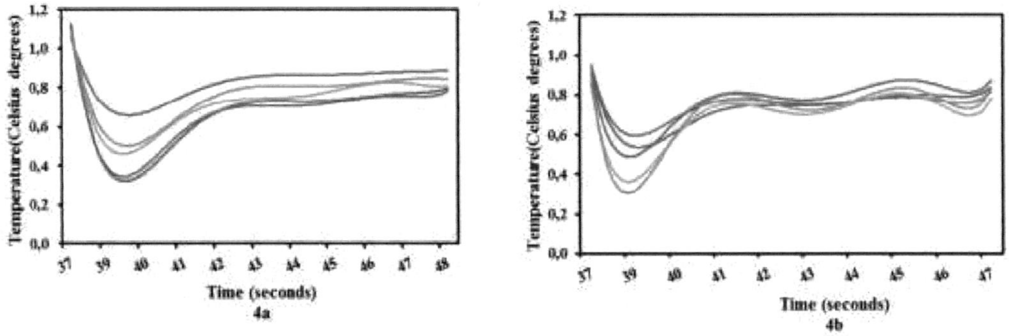

Figure 4. Thermograms of various spots a) in vertical and b) in horizontal position of foot. Ice cube was positioned over the right foot skin for about 40 second and then removed.

Conclusions

A typical infrared camera was used for a dynamic thermographic study of the lower limp in relation to position (vertical and horizontal). Return to thermal balance after initial cooling of a foot area followed roughly the same pattern for all points of measurement, although the heating rate was faster when the foot was in horizontal position. Thermal convection through blood flow appears to have a significant influence on skin temperature distribution. Gravity opposition to blood flow influences heat transfer mechanisms through blood flow.

4. References

[1] Domoina Ratovoson, Franck Jourdan and Vincent Huon 2013 Influence of gravity on the skin thermal behavior: experimental study using dynamic infrared thermography *Skin Research and Technology* **19** 397-408

[2] S. Bagavathiappan, B.B. Lahiri, T. Saravanan, John Philip , T. Jayakumar 2013 Infrared thermography for condition monitoring – A review *Infrared Physics & Technology* **55** 221-235

[3] E. Salazar-López, ,E. Domínguez , V. Juárez Ramos , J. de la Fuente, A. Meins, O. Iborra , G. Gálvez, M.A. Rodríguez-Artacho , E. Gómez-Milán July 2015 The mental and subjective skin: Emotion, empathy, feelings and thermography *Consciousness and Cognition* **34** 149–162

[4] Ismael Fernández-Cuevasa , Joao Carlos Bouzas Marinsb , Javier Arnáiz Lastrasa, , Pedro María Gómez Carmonaa, Sergio Piñonosa Canoa, , Miguel Ángel García-Concepcióna, , Manuel Sillero-Quintanaa 2015 Classification of factors influencing the use of infrared thermography in humans: A review *Infrared Physics & Technology* **71** 28–55

[5] A.S. Skouroliakou , I. Seferis , I. Sianoudis , I. Valais , A.F. Fragopoulou , L.H. Margaritis 2014 Infrared Thermography Imaging: Evaluating surface emissivity and skin thermal response to IR heating *e-Journal of Science & Technology (e-JST)* 9-14

Digital mammography with synchrotron radiation: characterization of a novel computed radiography system

Trivellato S[1], Vandenbroucke D[2], Arfelli F[3], Bessem M[2], Fedon C[3], Longo R[3], Tromba G[4], and Taibi A[1]

[1] Dipartimento di Fisica e Scienze della Terra, Università di Ferrara, via Saragat 1, 44122 Ferrara, Italy
[2] Agfa HealthCare NV, Septestraat 27, 2640 Mortsel, Belgium
[3] Dipartimento di Fisica, Università di Trieste, via Valerio 2, 34127 Trieste, Italy
[4] Elettra - Sincrotrone Trieste SCpA, SS 14 - km 163.5, 34149 Basovizza, Trieste, Italy

E-mail: taibi@fe.infn.it

Abstract. Breast X-ray imaging is a continuous research field to define dedicated equipment, with specialized X-ray sources and efficient detectors to improve image quality with an equal or even lower patient dose. The Needle Imaging Plate HM5.0, produced by Agfa, has been characterized using synchrotron radiation to assess the performance of this novel imaging chain in comparison to conventional mammographic equipment. The detection performance has been initially assessed in terms of Detective Quantum Efficiency (DQE) and its computation showed that DQE curves are very close to the typical results for digital radiography systems. Image threshold contrast has been then evaluated using the CDMAM phantom. The analysis has been completed with a scoring of visible details in the radiographs of the TORMAM phantom. The characterization thus confirms that monochromaticity leads to an equal image quality with a lower glandular dose and phase-contrast effects lead to an increase in anatomical structure detectability. Finally, a preliminary evaluation of clinical images showed a clear improvement in image quality thanks to phase-contrast contribution and to detector performance.

1. Introduction

It is well known that mammography is one of the most challenging techniques in medical imaging because of the low contrast between healthy and cancer tissues, high breast sensitivity to radiation effects and lesion conspicuity issues [1, 2]. Therefore, breast X-ray imaging is a continuous research field to improve image quality with an equal or even lower patient dose thanks to dedicated equipment, with specialized X-ray sources and efficient detectors.

The proposed imaging system has been designed to take advantage of the unique characteristics of its components: the Synchrotron Radiation (SR) and the detection performance of a novel Computed Radiography (CR) system. SR is considered an ideal X-ray source thanks to its monochromaticity and high spatial coherence. Monochromaticity helps to reduce patient dose avoiding low-energy components of conventional X-ray tubes. Furthermore, the high spatial coherence of the synchrotron beam allows the observation of phase-effects and the implementation of Phase-Contrast Imaging (PCI) [3, 4, 5]. Finally, the chosen device is a Needle Imaging Plate (NIP), produced by Agfa since 2010 (model HM5.0), based on the use of new phosphor material, the $CsBr:Eu^{2+}$. Detection efficiency and spatial resolution are improved with respect to the widespread Powder Imaging Plates (PIP) because of different structure and production method [6].

The analysis has been based on the comparison of the novel system with a clinical Full-Field Digital Mammography (FFDM) system available at the Cattinara Hospital in Trieste (Italy). The first part of the work consists in a quantitative evaluation of the detector performance according to the IEC standard. In the second part, Contrast-Detail (CD) performance and phase-contrast evidences have been analysed through the imaging of different test objects (mammography phantoms). Finally, the observed improvements have been verified in the clinical setting by analysing breast tissue images.

2. Description of the imaging chain

Radiographic images have been recorded at the ELETTRA Synchrotron facility located near Trieste. One of ELETTRA bending magnets is devoted to the SYRMEP collaboration (SYnchrotron Radiation for MEdical Physics), producing a flux of about 10^8 photons s^{-1} mm^{-2} [4]. The source is positioned about 30 m far from the sample to assure a high degree of spatial coherence and high collimation. At the SYRMEP beamline, the implemented PCI technique is the Propagation-based Phase-Contrast Imaging (PPCI) [3]. In particular, in the medical examination room a patient accommodation system allows the scanning of the patient's breast with the SR laminar beam [7]. The detection device is located in an anti-scatter holder which is placed on a 2 m-long rail. The detector can be thus translated to allow maximization of phase-contrast effects [4]. Previous studies at ELETTRA have been performed using screen-film and Dual-Side Imaging Plates [8]. The novel Agfa HM5.0 CR system uses a photostimulable phosphor deposited by Physical Vapour Deposition (PVD). In certain temperature and pressure conditions, phosphor crystals grow as needles. The used phosphor, $CsBr{:}Eu^{2+}$, is deposited with a density of 4.5 g cm^{-3}, an effective thickness of 110 μm and a 90% fill factor [9].

3. Analysis of detection system performance

The figure of merit suggested by the IEC standard is the Detective Quantum Efficiency (DQE) [10]. It is obtained as a function of the spatial frequency u from the Modulation Transfer Function *MTF(u)* and the Normalized Noise Power Spectrum *NNPS(u)*:

$$\mathrm{DQE}(u) = \frac{\mathrm{MTF}^2(u)}{q_0 \cdot \mathrm{K} \cdot \mathrm{NNPS}(u)}, \tag{1}$$

where q_0 is the number of photons per unit air kerma per mm^2 for the pertinent spectrum and K is the air kerma at the detector surface [9].

The resulting *DQE(u)* curves have been compared with the ones obtained with a conventional X-ray tube for mammography. Figure 1 shows that the detector performance is not influenced by the use of SR. Finally, it is worth noting that the Agfa HM5.0 shows detection performance comparable to state-of-the-art Digital Radiography (DR) systems, as reported by Marshall *et al* [9].

4. Analysis of Contrast-Detail (CD) Performance

CD performance of the novel imaging chain has been investigated with the CDMAM phantom (version 3.4, Artinis Medical System, The Netherlands) as suggested in the European Guidelines for Quality Assurance in Breast Cancer Screening and Diagnosis [11]. The limited horizontal dimension of the synchrotron beam does not allow to include the whole phantom surface. This prevents us from using the automatized CDCOM reader and human observations need to be performed following a common protocol. Figure 2 shows a comparison between a FFDM system manufactured by GE Healthcare and the analysed imaging chain. Although delivering a similar

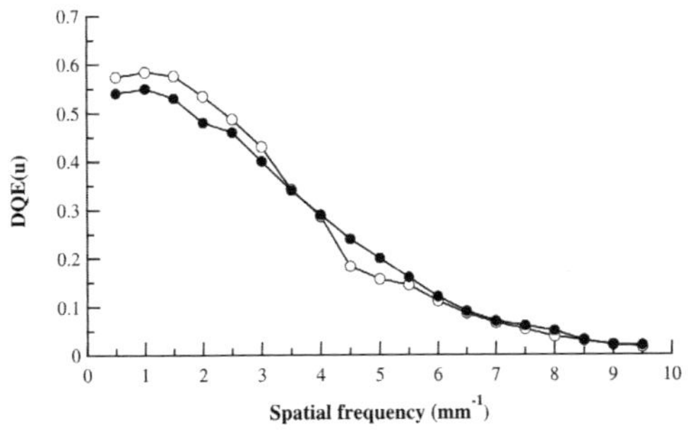

Figure 1: Horizontal DQE for the Agfa HM5.0 detector. Comparison between the use of SR (○) and the use of a conventional X-ray tube (●). Entrance air kerma values are 61 μGy and 77 μGy, respectively.

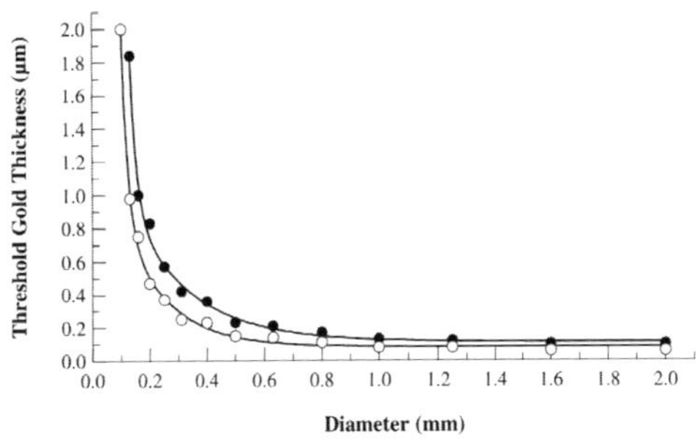

Figure 2: CD curve comparison. Data from a FFDM system have been acquired with a Rh/Rh tube, 29 kV, 50 mAs and a MGD of 1.16 mGy (●). SR data have been acquired at 18 keV and a MGD of 1.24 mGy (○).

Mean Glandular Dose (MGD), the novel imaging chain shows improved CD performance with a reduced mean threshold gold thickness. Explanations can be found in the detection performance, in the reduction of the scatter component and in the monochromaticity of the SR beam.

5. Phase-effect evidence in TORMAM phantom images

Images of TORMAM phantom (Leeds test objects, UK) have been acquired to compare scorings of mammography pertinent details mimicking spicules, microcalcifications and tumor masses. An example of the obtained scorings from a FFDM system and the novel imaging chain is reported in Table 1. The use of the novel imaging chain leads to an increased number of visible embedded details accompanied by the reduction of the delivered MGD. Furthermore phase-contrast significantly contributes to image formation leading to filaments detectability.

Table 1: TORMAM scoring for a phantom total thickness of 60 mm. The MGD values delivered with the FFDM and the SR systems are 3.00 mGy and 1.69 mGy, respectively.

Imaging system	Filaments	Microcalcifications	Masses
FFDM	0	1	4
SR	6	3	14

6. Evaluation of clinical images

A preliminary study of clinical images has been also performed by analysing mastectomy tissue radiographs kindly provided by the Cattinara Hospital in Trieste. As an example, an image comparisons is shown in Figure 3. A significant improvement of the overall image quality is apparent in the mammogram obtained with the novel imaging chain. As an average, an almost double contrast has been measured in correspondence of the microcalcifications imaged with the novel imaging chain.

7. Conclusions

The characterization of the novel imaging chain has highlighted that the joint use of SR with the detection performance of the Agfa HM5.0 system leads to a high quality mammography and it can expand the potential of phase-contrast imaging. TORMAM analysis investigated

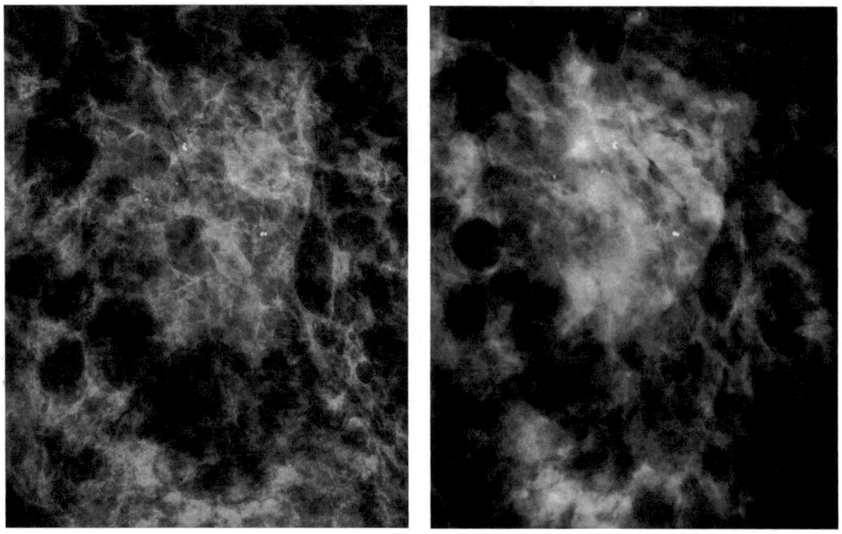

Figure 3: Tissue image comparison. On the left: image obtained with the novel imaging chain (16 keV, compressed thickness = 25 mm, MGD = 0.17 mGy). On the right: image obtained with a conventional mammography system (Mo/Mo, 23 kV, compressed thickness = 23 mm, MGD = 0.14 mGy).

the causes of higher Contrast-Detail performance: monochromaticity and phase-effects. The complete characterization has been then complemented with the radiography of breast specimen. A general improvement of image quality leads to a higher definition of the whole parenchymal structure, a confirmation of phase-contrast effectiveness.

Finally, this preliminary study demonstrated that improvements of diagnostic potential at minimum dose levels are still possible in breast X-ray imaging thanks to the latest technology developments.

Acknowledgements
The authors wish to thank Ingrid Berden (AGFA, Belgium) and Luigi Rigon (University of Trieste, Italy) for their help during the experimental work and the data analysis.

8. References

[1] Taibi A 2009 *Eur. J. Radiol.* **72** 447-53
[2] Taibi A and Vecchio S, *Breast Imaging.* In Brahme A (ed.) *Comprehensive Biomedical Physics* **2**, Amsterdam: Elsevier, 2014. 121-54
[3] Longo R, Tonutti M, Rigon L, Arfelli F, Dreossi D, Quai E, Zanconati F, Castelli E, Tromba G and Cova M A 2014 *Phil. Trans. R. Soc.* A **372**
[4] Dreossi D *et al* 2008 *Eur. J. Radiol.* **68** S58-S62
[5] Arfelli F *et al* 2000 *Radiology* **215** 286-93
[6] Leblans P, Vandenbrouke D and Willems P 2011 *Materials* **4** 1034-86
[7] Castelli E *et al* 2011 *Radiology* **259** 684-94
[8] Quai E, Longo R, Zanconati F, Jaconelli G, Tonutti M, Abrami A, Arfelli F, Dreossi D, Tromba G, Cova M A 2013 *Radiol. Med.* **118** 89-100
[9] Marshall N W, Lemmens K and Bosmans H 2012 *Med. Phys.* **39** 811-24
[10] IEC 62220-1-2 2005 *Medical electrical equipment - Characteristics of digital X-ray image devices - Part I: Determination of the detective quantum efficiency* (International Electrotechnical Commission, Switzerland)
[11] EC Guidelines, In: Perry N, Broeders M, de Wolf C, Tornberg S, Holland R, Von Karsa L, and Puthaar E (eds.) 2006, *European Guidelines for Quality Assurance in Breast Cancer Screening and Diagnosis*, 4th edn., Luxembourg: Office for Official Publications of the European Communities

THE APPLICATION OF COMPUTED TOMOGRAPHY IN WOUND BALLISTICS RESEARCH

Nick Tsiatis[1], Konstantinos Moraitis[1], Stavroula Papadodima[1], Chara Spiliopoulou[1], Alexis Kelekis[2], Christos Kelesis[2], Efstathios Efstathopoulos[2], Sofia Kordolaimi[2], Agapi Ploussi[2]

[1]Department of Forensic Medicine and Toxicology, School of Medicine, University of Athens, Greece.
[2]2nd Department of Radiology, Medical School, University of Athens, Greece.
Corresponding Author: Nick Tsiatis, tsiatisnick@yahoo.gr

Keywords: wound ballistics, human body, density, CT, Hounsfield.

Abstract

In wound ballistics research there is a relationship between the data that characterize a bullet and the injury resulted after shooting when it perforates the human body. The bullet path in the human body following skin perforation as well as the damaging effect cannot always be predictable as they depend on various factors such as the bullet's characteristics (velocity, distance, type of firearm and so on) and the tissue types that the bullet passes through. The purpose of this presentation is to highlight the contribution of Computed Tomography (CT) in wound ballistics research. Using CT technology and studying virtual "slices" of specific areas on scanned human bodies, allows the evaluation of density and thickness of the skin, the subcutaneous tissue, the muscles, the vital organs and the bones. Density data taken from Hounsfield units can be converted in g/ml by using the appropriate software. By evaluating the results of this study, the anatomy of the human body utilizing ballistic gel will be reproduced in order to simulate the path that a bullet follows. The biophysical analysis in wound ballistics provides another application of CT technology, which is commonly used for diagnostic and therapeutic purposes in various medical disciplines.

INTRODUCTION

The majority of firearms follow a similar way of operation which is consisted in a particular process for any type of them. After pulling the trigger an explosion takes place while the primer of the ammunition is hit, a chemical reaction creates expanded gases through which the bullet is detached and moves in the internal of the guns` barrel accelerating to exit from the muzzle. Then it follows a linear and simultaneously a rotational motion against the external conditions of the environment. This movement depends on various factors such as the muzzle velocity, the caliber, the mass, its geometrical characteristics (tip, heel and ogive shape) and the type of the used firearm.

Besides, according the bullets` trajectory, the injury resulted when a bullet perforates the human body depends on the distance, the angle of impact and the part of the body through which it enters. The bullet path in the human body after its skin perforation cannot always be predictable. The extent of injury depends on various factors such as the type of the firearm and its ammunition, its angular momentum, the shape, the caliber, etc. In addition, the trajectory has an effect on the path that the bullet will follow inside the human body in combination with the muscles and the vital organs that could be hit.

When a bullet hits the human body, it creates an entrance wound on the point of impact, the shape of which depends mainly on the angle of incidence and the distance of shooting.

Human body simulation

Experimental shootings on materials similar to human body are necessary to be performed. Specifically, the ballistic soap and the ballistic gel, which can be produced under concrete conditions, enable us to write down the path that a bullet follows as a simulation in a human body. Nevertheless, it is important to recognize that the best of measuremnts is an estimate and has some uncertainty associated with it. Based on numerous test shots into tissue stimulants, an uncertainty of $\pm5^{o}$ to $\pm10^{o}$ for wound paths is not unrealistic [1]. We can also notice the temporary and the permanent cavity that are caused by the energy deposited when the bullet passes through. It is very important to record the influence on human body that different types of bullets have. The scope through the CT technology is initially the calculation of density in different areas of the human body; We can then produce the ballistic gel for more accurate simulation.

The "Ballistic gelatine" is a protein produced by submitting collagen to an irreversible process that renders it water-soluble. It is usually made of skin or bone. The raw materials and the manufacturing process influence its characteristics like molecular weight, isoelectric point, transmittance, viscosity and gel strength. The last two parameters are of particular relevance to the use of gelatin for modeling. Gelatin is obtained by treatment with acid or alkali, followed by extraction using hot water. Gelatin is amphoteric; it reacts to an electrical field like a cation in an acidic solution and like an anion in an alkaline solution. At a particular pH value, the reaction of the solution is neutral. This is known as the isoelectric point. The pH value of the isoelectric point depends highly on the treatment to which the collagen has been subjected. Treatment with alkali produces a Type B gelatin, with an isoelectric point somewhere between pH 4.7 and 5.4. The gel strength of the gelatin (which is of importance for its strength) is measured in Bloom[2].

The "Ballistic soap" that can be used for wound ballistics experiments is glycerine soap (moulded, transparent soap). It can be produced by mixing the constituent fats and oils with care, it is possible to maintain consistency over periods of many years, which is an important prerequisite for accuracy. It is possible to modify the density and softness of soap by varying its components. This allows us to achieve the penetration depth required for a particular calibration shot. It is therefore perfectly possible to standardize ballistic soap.

Depending on its composition, the density of new soap (1 to 3 weeks old) lies between 1060 and 1100 kg/m^3 at room temperature. Soap has a very low limit of elasticity (0.5 N/mm^2), which is an indicator of its plastic behavior. The deformation of soap due to the passage of a bullet is almost entirely plastic, and the wound channel only collapses very slightly. This gives us a direct image of the temporary cavity, on which we can readily take measurements[2]. Using the CT technology the thickness of layers in several areas of the human body can be calculated so as to simulate the wound channels.

PROPOSED METHODOLOGY

The aim is to highlight the contribution of Computed Tomography in wound ballistics research. Using CT technology and studying virtual "slices" of specific areas on scanned human bodies allows the evaluation of density and thickness of the skin, the subcutaneous tissue, the muscles, the vital organs and the bones. Density data taken from Hounsfield units can be converted in g/ml by using the appropriate software.

Data are collected by using DICOM images from people of different age, sex and weight that underwent CT examinations on a Philips Brilliance 64-slice CT scanner installed at the 2nd Department of Radiology in University General Hospital ATTIKON (Athens, Greece); similar ones are mentioned by Dr. Les Folio as they had been used for many radiology exams[3]. Data are rewritten following the rules imposed by the Bioethics Commission of the School of Medicine. Cross section slices have been collected from many areas of scanned human bodies (from patients who have been hospitalized in "ATTIKON" for various reasons), through the "Evorad RIS-PACS 3.0.0. build 170" (Evolving information technology for Radiology) software for visual images. Measurements are still taken for the thickness of the layers and for their density. In order to achieve satisfying statistical analysis results, following the biostatistics standards, the aim is to manage collecting data from 300 individuals - patients (males and females). The population sample should consist of two groups: the first includes individuals from twenty to fifty (20-50) years old and a second one includes individuals from fifty one and over (51+) years old.

An additional important parameter should be the weight, which means that each one of these two groups includes 3 weight categories: underweight (30–60 kgrs), normal weight (61–90 kgrs) and overweight ones (91+).
The areas from the human body have been chosen in order to cover almost the whole body surface, except the head (brain) and the area under the knees. This happens because there are no sufficient data from whole body examinations where total area of the patients` body (from the top head to the toes) is scanned.
Besides, for the sake of wound ballistics research, the human body's figure has been "separated" in five parts according to the morphology, the muscles` mass division, the shape and the bones` form, as following: head - brain, arms - forearms, thorax - chest, abdomen and foot.

The cross section slices that have been chosen for taking measurements should have been marked keeping the same reference point for each computed tomography which is being examined as a new sample. For this purpose these slices cover the following areas: a) shoulder girdle, b) lungs and heart, c) liver, spleen and stomach, d) kidneys, intestine and pancreas, e) pelvis and forearms, f) middle of thigh.[4]

Measurements

For each cross section slice measurements for the density and the thickness from all the layers are taken, in particular measurements conducted in the skin, the subcutaneous tissue, muscles, vessels and the vital organs included in these slices (e.g. lungs, heart, liver, stomach, etc.). The region of interest (ROI) covers a surface of 50 - 55mm^2, except the skin where it is not possible the ROI to be more than 20mm^2, because of its oblong shape (Fig. 1 & 2). Furthermore, the choice of the ROI could be a potential source of error as it is nearly impossible to eliminate the human error factor in selecting the ROI [5]. Moreover, standardizing the ROI could be difficult because of the anatomical differences among organs. Density is recorded in Hounsfield units and the standard deviation (SD) is also received for each measurement.

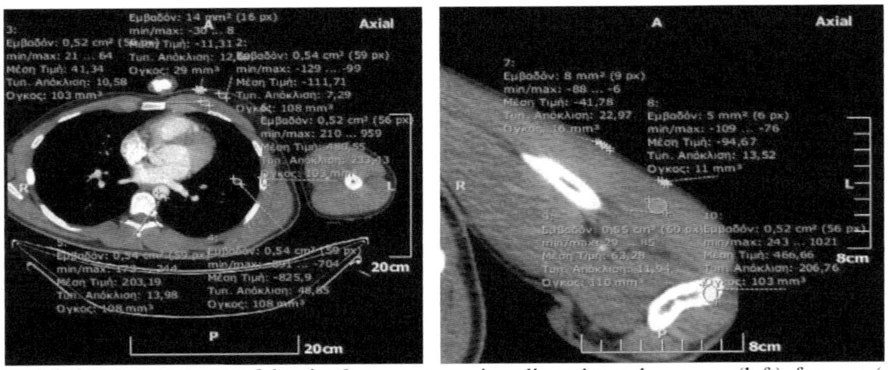

Figures 1 & 2: Measurements of density from cross section slices: lungs-heart area (left), forearm (right).

Each pixel is assigned a numerical value (CT number), that is compared to the attenuation value of water and displayed on a scale of arbitrary units named Hounsfield units (HU). This scale assigns water as an attenuation value (HU) of zero. For a range of CT numbers in 2000 HU wide, each numeral value represents a shade of grey with +1000 (white) and –1000 (black) at either end of the spectrum (although modern scanners have a greater range of HU up to 4000).

The thickness is written down in cm (Fig. 3 & 4). All the values are collected in tables separately for each area of the human body and their recording follows the age group where the tomography correlates. Each age group includes the three categories of weight. The next step is to convert the Hounsfield units to units of density (gr/cm^3), before calculating the average values for each area and layer.

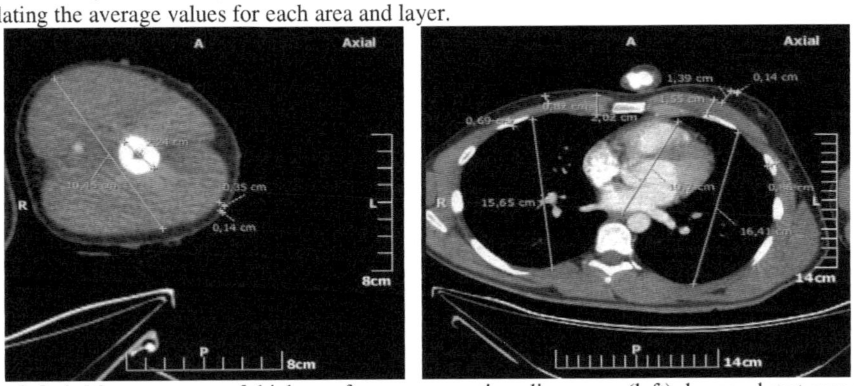

Figures 3 & 4: Measurements of thickness from cross section slices: arm (left), lungs – heart area (right).

Preliminary results

To convert the HU units to units of mass density (gr/cm^3) the Matlab code 'hounsfield2density' is used[6], [7], (process which is in progress). In the following table there are indicative measurements of the density from the area of thorax.

MEASUREMENTS OF DENSITY IN LAYERS OF HUMAN BODY MALES-FEMALES / AGE: 20-50 yrs / Normal Weight / AREA: Thorax / Mass Density Units-gr/cm^3													
SKIN	SD	SUBCUTANEOUS	SD	MUSCLES	SD	BONES	SD	LUNG	SD	LIVER	SD	HEART	SD
0,974	0,0043	0,894793062	0,0043	1,051469	0,0052	1,2029	0,093	0,1707	0,0438	1,06938	0,00734	1,06319	0,00683
1,057	0,0469	0,918667831	0,0241	1,056473	0,0227	1,1499	0,166	0,1639	0,0297	1,05423	0,03318	1,04235	0,02132
0,958	0,0056	0,885676736	0,0079	1,046073	0,0136	1,1956	0,07	0,1792	0,0098	1,08183	0,00886	1,0963	0,00758
0,99	0,0072	0,894094178	0,0017	1,037573	0,005	1,3175	0,231	0,16	0,0234				
1,139	0,0686	0,920096063	0,0256	1,047923	0,0327	1,1876	0,286	0,1551	0,0588				
0,924	0,0018	0,892615303	0,0053	1,044802	0,003	1,1916	0,18	0,2237	0,0581				
0,992	0,0046	0,908781648	0,0063	1,044281	0,0075	1,2033	0,016						

1,079	0,0913	0,931268622	0,0396	1,049448	0,029	1,1693	0,051						
1,001	0,0078	0,910250224	0,01	1,049887	0,0071	1,1778	0,113						
1,013±0,0265		**0,906249296±0,0139**		**1,047548±0,014**		**1,1995±0,134**		**0,1754±0,0373**		**1,06848±0,01646**		**1,06728±0,01191**	

Table 1: Measurements of density (gr/cm^3) performed from three patients (males) after CTs (numbers in bold show the average values).

DENSITY IN CROSS SECTION AREAS OF HUMAN BODY MALES-FEMALES / AGE: 20-50 yrs / Normal Weight								
AREA	THORAX		ABDOMEN		ARMS-FOREARMS		THIGH	
(gr/cm^3)	DENSITY	SD	DENSITY	SD	DENSITY	SD	DENSITY	SD
	1,012792	0,026461	1,008	0,0378	0,989	0,0251	0,954	0,0185
	0,906249	0,013857	0,906217	0,0121	0,925172	0,0173	0,908994	0,0065
	1,047548	0,01398	1,05191	0,01	1,053772	0,0131	1,061856	0,0083
	1,19948	0,13405	1,2901	0,1653	1,511	0,177	1,5519	0,192
	0,17544	0,037251	0,35299	0,01536				
	1,002307	0,010932	1,06723	0,01834				
	1,06848	0,016455	1,095201	0,019788				
	1,06728	0,011911	1,104355	0,018579				
			1,091729	0,006767				
			1,091182	0,010749				
Average	**0,934947**±0,033112		**1,005891**±0,031478		**1,119188**±0,056325		**1,119736**±0,058125	

Table 2: Density (gr/cm^3) for the areas of thorax, abdomen, arms-forearms & thigh from normal weight patients.

CONCLUSIONS

The biophysical analysis in wound ballistics provides a further application of CT, which is commonly used for diagnostic and therapeutic purposes in various medical disciplines. The ability in taking measurements of the density and the thickness from the layers of human bodies gives the opportunity to calculate values on average, especially to the density according to which ballistic gelatine can be produced for more pragmatic and accurate results. Through experimental shootings in these materials that simulate human body, we can realize and approach the estimated performance of bullets after perforating different parts of human body. This is a study in progress and the final results are expected to be published soon.

ACKNOWLEDGMENTS

I feel obliged to express my thanksgiving to Mr Haag Lucien C. (Forensic Science Services), for his advice.

REFERENCES

[1] Haag L. C., *Shooting Incident Reconstruction*, Elsevier, Amsterdam, 2006, ISBN 13:978-0-12-088473-5.

[2] Kneubuehl, Beat P., Robin M. Coupland, Markus A. Rothschild, Michael J. Thali, *Wound Ballistics-Basics and Applications*, Translation of the revised third German edition Springer-Verlag Berlin Heidelberg New York, 2008, ISBN-13 978-3-642-20355-8.

[3] Folio L., *Combat Radiology Diagnostic Imaging of Blast and Ballistic Injuries*, Springer, New York Heidelberg Dordrecht London, 2010, ISBN 978 1 4419 5853 2.

[4] Gosling J.A., Harris P.F., Whitmore I., Willan P.L.T., *Human Anatomy-Color Atlas and Text* – Translated from Anagnostopoulou Sofia, 4th Edition, Scientific Publication Parisianou Ltd., 2003, ISBN 960-394-226-X.

[5] Isabela Maria de Carvalho Crusoé Silva, Deborah Queiroz de Freitas, Glaucia Maria Bovi Ambrosano, Frab Norberto Bóscolo, Solange Maria Almeida, *Bone density: comparative evaluation of Hounsfield units in multislice and cone-beam computed tomography*, Braz Oral Res. (Sao Paulo), 6, 550-556, 2012.

[6] Schneider U., Pedroni E., and Lomax A., *The calibration of CT Hounsfield units for radiotherapy treatment planning,* Phys. Med. Biol., 41, 111-124, 1996.

[7] http://www.k-wave.org (accessed 2015 March).

Incorporation of a two metre long PET scanner in STIR

C Tsoumpas†[1], C Brain[2], T Dyke[2], D Gold[1]
[1] Division of Biomedical Imaging, University of Leeds, Leeds, LS2 9JT, UK
[2] School of Physics and Astronomy, University of Leeds, Leeds, LS2 9JT, UK

E-mail: c.tsoumpas@leeds.ac.uk

Abstract. The Explorer project aims to investigate the potential benefits of a total-body 2 metre long PET scanner. The following investigation incorporates this scanner in STIR library and demonstrates the capabilities and weaknesses of existing reconstruction (FBP and OSEM) and single scatter simulation algorithms. It was found that sensible images are reconstructed but at the expense of high memory and processing time demands. FBP requires 4 hours on a core; OSEM: 2 hours per iteration if ran in parallel on 15-cores of a high performance computer. The single scatter simulation algorithm shows that on a short scale, up to a fifth of the scanner length, the assumption that the scatter between direct rings is similar to the scatter between the oblique rings is approximately valid. However, for more extreme cases this assumption is not longer valid, which illustrates that consideration of the oblique rings within the single scatter simulation will be necessary, if this scatter correction is the method of choice.

1. Introduction

Data acquisition for Positron Emission Tomography (PET) produces three-dimensional images but only at a limited axial field of view (approximately 20cm). Currently in order to perform full body PET, multiple images at different bed positions are acquired [1]. The pioneering 'Explorer Project' proposes that a total body PET scanner needs to be built so that the entire human body can be imaged at once. The 'Explorer Project' is an international consortium of scientists developing a 2m long PET scanner: http://explorer.ucdavis.edu/. The predicted effective sensitivity of such a scanner will be 40-fold higher than current scanners. This can result in as low radiation dose as 150 μSv, which could have huge implications for the justification of imaging humans even at healthy conditions or at prenatal phase [2]. The technology is currently being developed and image reconstruction techniques need to be advanced in order to process the large amount of data and produce the highest quality images possible.

2. Aim

This project aims to provide a preliminary evaluation of reconstruction and correction methods of the open access Software for Tomographic Image Reconstruction (STIR) for the total-body PET scanner [3]. First we assess the reconstruction settings and computational times for this scanner. Second we explore the extent of validity of using the current scheme of the single scatter simulation (SSS) algorithm for scatter correction, which assumes that scatter between indirect rings is similar to scatter within the same ring.

† To whom any correspondence should be addressed.

3. Materials and Methods

All simulations and reconstructions of phantom data were performed using the STIR library (release 3.0) on a 3040-core high performance computer at the University of Leeds (ARC2): 8-core Intel E5-2670 2.6 GHz processors. Each job has access to 32 GB of memory.

3.1. Phantom

We employed the XCAT 2.0 total-body anthropomorphic phantom [4], which assigned realistic F18-FDG PET values as well as representative attenuation values to serve as input images for the simulation. In this experimental setup both images consisted of 195×195×1150 voxels with sizes 3.42×3.42×1.71 mm^3.

3.2. Total-body scanner

The simulated total-body PET system consists of 36 block rings and 48 block modules per ring in which each detector block consisted of a 15×15 array of 3.42×3.42×20 mm^3 LSO crystals [5]. The inner diameter of the system is 80 cm corresponding to a 70 cm trans-axial FOV. The axial FOV is 196 cm with one crystal gap (3.42 cm) between adjacent rings. We set the energy window as 440 keV to 665 keV and energy resolution as 13%. The corresponding parameters that are incorporated in STIR are provided in the Appendix section.

3.2.1. Simulation of attenuated line integrals

The emission and attenuation sinograms were estimated using the corresponding forward-projection STIR functions and the attenuated emission data were produced. Poisson noise was inserted to empirically simulate two noise levels: (A) 500 million counts of data (no randoms or scatter), which are expected to be obtained from a 20 min F18-FDG scan (10 MBq total activity); (B) approximately 2.5 billion counts (no randoms or scatter), which resembles a scan with higher activity. The forward projection and calculation of the attenuation factors required approximately 5.5 hours. We used a number of compressions to reduce memory demands: Span: 11, Max ring difference: 336, Mash: 2, Rays tracing per bin: 1 transaxially × 2 or 3 axially.

3.2.2. Scatter simulation. Running the SSS algorithm in STIR requires substantial down sampling of the scanner detectors in order to achieve reasonable computational times and memory requirements [6]. The number of total rings was reduced by a factor of 25: from 575 to 23, maintaining the length of the scanner. SSS was performed for all ring differences to assess how the scatter profile depends on the ring difference. More details on the scatter simulation parameters are provided in the Appendix. The attenuation map was down-sampled to produce the scatter-points map: 29×29×288 voxels with sizes 24.4×24.4×6.84 mm^3. The fully 3D SSS used 24,268 scatter points and required 50 minutes and 680 MB of memory.

3.3. Reconstruction

The reconstruction algorithms that have been selected for this study are: filtered back projection with 3D reprojection (FBP-3DRP with defaults settings for any scanner as in STIR: release 3.0) and attenuation pre-corrected 3D OSEM (8 iterations, 15 subsets). The computational time and memory for FBP-3DRP was approximately 4 hours and 500 MB. 3D OSEM on a single core required 9 hours per full iteration and 5 GB of memory, while when ran using the parallel message passage interface (MPI) implementation of OSEM on 15 cores it required a little more than 2 hours per full iteration and 15 GB of memory. The reconstructed images have the same size as the simulated images.

Journal of Physics: Conference Series **637** (2015) 012030 doi:10.1088/1742-6596/637/1/012030

4. Results

Figure 1 demonstrates coronal view of the reconstructed whole body simulated F18-FDG images using FBP-3DRP and OSEM for both noise levels. Figure 2 demonstrates one slice of the single scatter probability sinograms between the same ring and one for ring difference 13 of the down sampled scanner detectors (i.e. equivalent to 325 ring difference for the original total-body scanner). Figure 3 provides summed line profiles through the central sinogram plane for each ring difference.

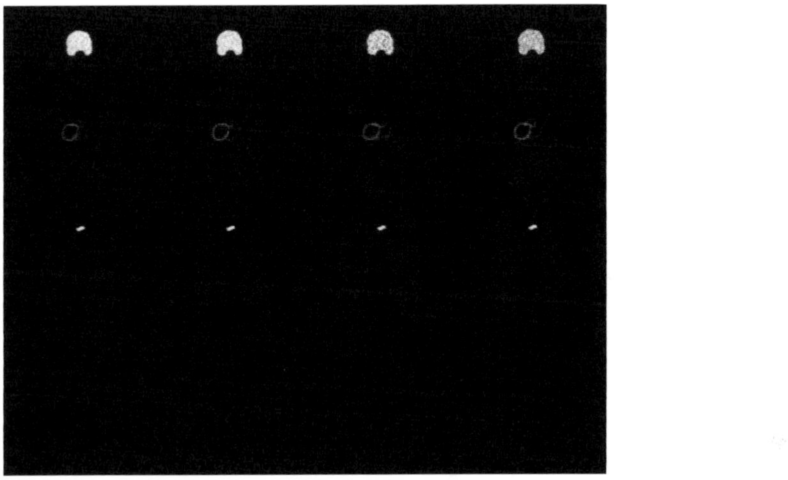

Figure 1: Coronal view of the total-body reconstructions. From left to right-hand side: FBP-3DRP 2.5×10^9 counts, OSEM 2.5×10^9 counts, FBP-3DRP 5×10^8 counts, OSEM 5×10^8 counts

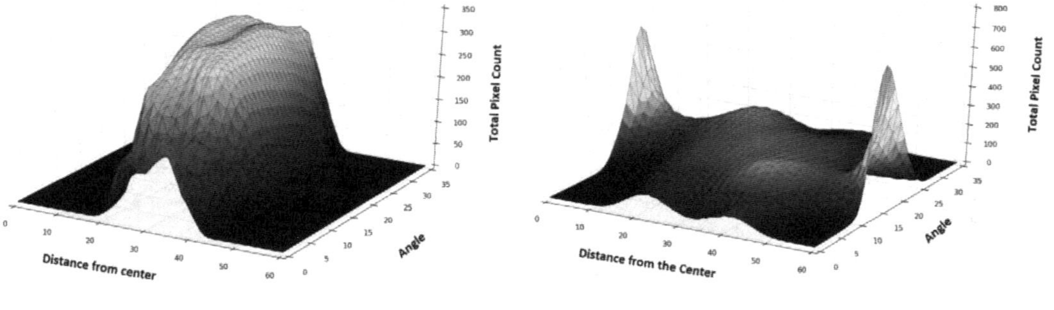

Figure 2: 3D plot of the scatter sinogram for direct rings (left-hand side) versus the scatter sinogram for thirteen 'compressed' ring difference (i.e. equivalent to 325 ring difference) sinogram (right-hand side)

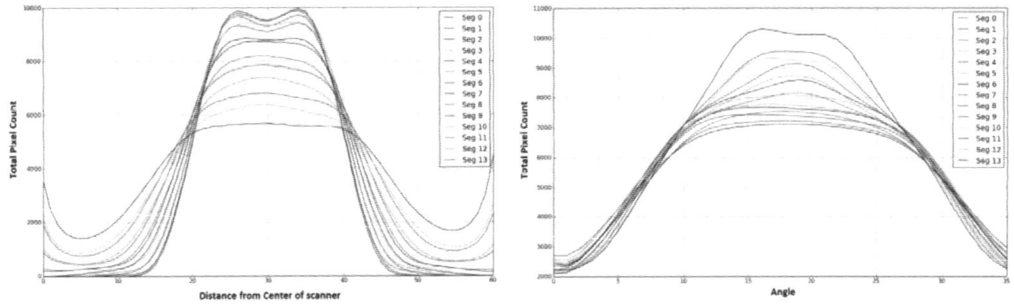

Figure 3: Profiles for 0 to 13 'compressed' ring difference: transaxial-direction profiles summed along angular-direction (left-hand side); angular-direction profiles summed along transaxial direction (right-hand side)

5. Discussion and Conclusion

The aim of this project was to test the reconstruction performance of STIR 3.0 library for a total-body system and investigate the validity of the current scatter simulation method for such a scanner. Through the simulations ran with the STIR library it was concluded from figures 2 and 3 that the method of approximating the scatter between oblique sinograms to the direct sinograms was imperfect, for large angles. For ring difference between 100 and 150, the discrepancy of scatter between direct and indirect rings is around 10%. Higher ring difference creates larger discrepancy and the accuracy of the reconstructed image is likely to be affected. Due to this it is clear that the scatter algorithm needs to be updated to deal with large oblique angles in the longer scanner.

Reconstruction of the total-body simulated data reflects the necessity of parallel computational processing for speeding up the computations. The analytical FBP-3DRP reconstruction algorithm for compressed sinogram data with span 11 and mash 2 requires about 4 hours to produce images, while the MPI version of the 3D OSEM reconstruction algorithm takes 2 hours per full iteration. Future work will focus on the parallel processing implementation of the projectors and scatter simulation algorithm. As the current version of interpolating the scatter sinograms will not provide valid results, appropriate oblique sinograms sampling and interpolation will be necessary, which will introduce more computing requirements. In addition, scatter simulation needs further exploration, as for example multiple scatter may be more significant for the total-body PET scanner. It may be the case that Monte Carlo based scatter correction methods is more appropriate for this scanner. Furthermore, incorporating additional information from time-of-flight, point-spread-function and depth-of-interaction will be essential to maximise the potential of this novel system. In order to include all this information, list-mode based iterative reconstruction will be more efficient. In conclusion, although the STIR library is capable in producing data that can be useful to researchers, it requires a series of major updates to cope with the large amount of data acquired from a total-body PET imaging system.

6. References

[1] Cherry S R 2006 *J. Nucl. Med.* **47** 1735
[2] Jones T and Budinger T F 2013 *J. Nucl. Med.* **54** 2016
[3] Thielemans K, Tsoumpas C, Mustafovic S, Beisel T, Aguiar P, Dikaios N and Jacobson M W 2012 *Phys. Med. Biol.* **57** 867
[4] Segars W P, Sturgeon G, Mendonca S, Grimes J and Tsui B M W 2010 *Med Phys,* **37** 4902
[5] Poon J K, Dahlbom M L, Moses W W, Balakrishnan K, Wang W, Cherry S R and Badawi R D 2012 *Phys Med Biol.* **57** 4077
[6] Polycarpou I, Thielemans K, Manjeshwar R, Aguiar P, Marsden P K and Tsoumpas C 2011 *Ann. Nucl. Med.* **25** 643

Acknowledgments

The authors wish to thank Xuezhu Zhang, Ramsey Badawi, Simon Cherry and Jinyi Qi (UC Davis) for providing information on the scanner design and the anatomical and functional representation of the phantom. Furthermore, we are grateful to the members of the Advanced Research Computing lab, University of Leeds and particularly Martin Callaghan for fruitful advice and suggestions on the HPC.

Appendix: Inserting the scanner in STIR

Part of the Scanner Information Configuration:

```
case UltraPET: set_params(UltraPET, string_list("UltraPET","UPET"), 575, 244, /* (15*36)
+35 */ 229,15*48, 400.F, 8.F, 3.42F, 3.42F, 0.F, 0, 0, 0, 0, 0, 0, 1);
```

Part of the Scatter Template Interfile Information:

```
Segments: 27; Views: 36; Axial coordinate range: From 10 to 23; Tangential coordinate:
61; Minimum & Maximum ring difference per segment: From  -13, to 13; Number of rings:
23; Number of detectors per ring: 72; Inner ring diameter: 80; Distance between rings
(cm): 8.55; Default bin size (cm): 1.368; Maximum number of non-arc-corrected bins: 61;
Default number of arc-corrected bins: 57; effective central bin size (cm): 0.712
```

Effect of the Concentration on the X-ray Luminescence Efficiency of a Cadmium Selenide/Zinc Sulfide (CdSe/ZnS) Quantum Dot Nanoparticle Solution

I Valais[1], C Michail[1], D Nikolopoulos[2], C Fountzoula[3], A Bakas[4], P Yannakopoulos[2], G Fountos[1], G Panayiotakis[5] and I Kandarakis[1]
[1]Radiation Physics, Materials Technology and Biomedical Imaging Laboratory, Department of Biomedical Engineering, Technological Educational Institute of Athens, Egaleo, 122 10 Athens, Greece
[2] Department of Computer Electronic Engineering, Piraeus University of Applied Sciences (Technological Education Institute of Piraeus), Greece, Petrou Ralli & Thivon 250, GR122 44, Aigaleo, Athens, Greece
[3] Department of Medical Laboratories, Technological Educational Institute of Athens, Egaleo, Athens, Greece
[4] Medical Radiological Technology, Faculty of Health and Caring Professions, Technological Educational Institute of Athens, Athens, Greece
[5]Department of Medical Physics, School of Medicine, University of Patras, 15310, Rion, Greece

e-mail: valais@teiath.gr

Abstract. In the current study preliminary results on the luminescence efficiency (LE) of toluene dissolved Cadmium Selenide/Zinc Sulfide (CdSe/ZnS, Sigma-Aldrich, Lumidot 694622) quantum dot samples (QDs) after exposure to X-rays of variable radiation flux are shown. The distinctive influence of the weight over volume (w/v) concentration of the samples in LE was investigated. The light emission of the QDs was additionally measured after UV irradiation. The distribution of the emitted light was symmetrical with a maximum at 590 nm. The w/v concentration of the QDs varied between $7.1x10^{-5}$ mg/mL to $28.4x10^{-5}$ mg/mL. The samples were handled in a cubic $12.5x12.5x45mm^3$ quartz cuvette. Each sample was excited under X-ray irradiation, in the energy range from 50 to 130 kVp using a BMI General Medical Merate tube with rotating Tungsten anode and inherent filtration equivalent to 2 mm Al. The X-ray LE, induced by the $28.4x10^{-5}$ mg/mL QDs found higher, however, the distinction was vague in the highly concentrated samples. The maximum efficiency was obtained at the 90 kVp for QDs with $21.3x10^{-5}$ mg/mL w/v concentration. In the high energy range (120-130 kVp) all concentration levels exhibited comparable X-ray induced LE. The luminescence properties of the investigated QDs appear promising for X-ray detection applications.

Keywords: Quantum Dots; Luminescence Efficiency; CdSe/ZnS

1. Introduction
X-ray medical imaging detectors incorporate a scintillator layer combined with various types of optical sensors [1-14]. Phosphors or scintillators, emit light upon ionizing radiation excitation (X, gamma etc.), which is captured by optical sensors. A development in scintillators is materials whith reduced grain size [7, 10, 15-17]. Nanophosphors materials with these properties are Quantum Dots (QDs). [18-24]. QDs are semiconductor nano-crystals with particle size 1-20nm. Due to their unique optical and electrical properties have already been applied in optoelectronic sensors [23-26]. In comparison with traditional scintillators, QDs exhibit better resolution, lower decay time and absence of afterglow, leading to faster response. Furthermore QDs have higher detection efficiency since they are prepared by high atomic number and density materials, such as CdTe, CdHgTe, PbSe and PbTe. Their band gap energy (Eg) can be adjusted in order to be optimally combined with the maximum spectral sensitivity of charged coupled devices (CCD) and complementary metal oxide

semiconductors (CMOS) [16,17,27,28]. However the reduced scattering, due to the reduced grain size, could downgrade image resolution due to high diffusion of emitted light. On the other hand this behavior may increase the emitted photons which reach the imaging device. The aim of the present study was to investigate the effect of the concentration of Cadmium Selenide/Zinc Sulfide (CdSe/ZnS) core-shell type quantum dot nanoparticles on the luminescence efficiency, in the radiography energy range.

2. Materials and Methods

2.1. Quantum Dots preparation

The CdSe/ZnS quantum dot samples were purchased from Sigma-Aldrich (Lumidot 694622) and were dissolved in toluene. CdSe/ZnS has particle size of 4 nm, extinction coefficient 1.6×10^5 $L \cdot mol^{-1} cm^{-1}$, density of 0.865 g/ml at 25^0, excitation wavelength (λ_{ex}) 575 nm, emitting wavelength 590 nm, full width at half maximum <40 nm and a quantum yield ≥30% [29]. The attenuation of light, at a given wavelength is described by the extinction coefficient. The particular CdSe/ZnS QD has an intermediate extinction coefficient value, compared to other QDs of the same series [29]. This was selected since others with very small extinction coefficient (0.02×10^5 $L \cdot mol^{-1} cm^{-1}$) emit light in the blue region of the optical spectrum, being incompatible with the most common digital optical sensors which are sensitive in the red part of the spectrum. The samples were handled in a cubic $12.5 \times 12.5 \times 45 mm^3$ quartz cuvette. The distinctive influence of the weight over volume (w/v) concentration of the samples in LE was investigated. The w/v concentration of the QDs varied between 7.1×10^{-5} mg/mL to 28.4×10^{-5} mg/mL. QDs were exposed to X-rays on a BMI General Medical Merate tube with rotating Tungsten anode and inherent filtration equivalent to 2 mm Al, with energies ranging from 50 to 130 kVp. An additional 20 mm filtration was introduced in the beam to simulate beam quality alternation by a human body [11].

2.2. Emission spectrum

The emitted light was measured by a grating optical spectrometer (Ocean Optics Inc., HR2000). The mean light photon energy $\overline{E\lambda}$ was determined from the emitted light spectrum of the CdSe/ZnS quantum dot samples.

2.3. Luminescence efficiency (LE)

The efficiency of a scintillator to emit light, after X-ray exposure, can be experimentally determined under clinical conditions by their luminescence efficiency (LE), defined in terms of emitted light energy flux $\dot{\Psi}_\lambda$ per unit of incident exposure rate, i.e.:

$$\eta_A = \dot{\Psi}_\lambda / \dot{X} \tag{1}$$

Where $\dot{X}$ is the exposure rate measured with a Piranha P100B (RTI) dosimeter. AE is expressed in efficiency units (E.U.) $\mu W \times m^{-2} / (mR \times s^{-1})$. The light flux measurements were performed using a light integration sphere (Oriel 70451), coupled to a photomultiplier (PMT) (EMI 9798B), connected to a Cary 401 vibrating reed electrometer [9].

3. Results and Discussion

Figure 1 shows the emitted optical spectrum of CdSe/ZnS quantum dot samples normalized to unity in the 400-700 nm wavelength region. The CdSe/ZnS quantum dot samples spectrum shows maximum at 590 nm, lying in the yellow region of the optical spectrum. Having a maximum at 590 nm, the mean light photon energy ($\overline{E\lambda} = hc / \overline{\lambda}$) results to 2.11 eV.

Figure 2 shows the variation of the luminescence efficiency of the CdSe/ZnS quantum dots with X-ray tube voltage, in the range from 50 to 130 kVp. Luminescence efficiency values of the quantum dots were increased with increasing concentration and saturated in the concentration range of 300 to 400 mg/μL. The maximum efficiency was obtained at 90 kVp for a QD concentration of 300 mg/μL. In the high energy range (120-130 kVp) all concentration levels exhibited close luminescence efficiency values.

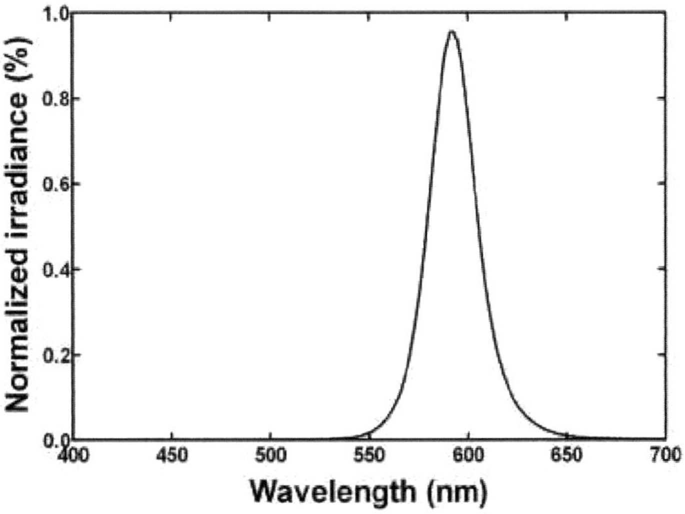

Figure 1. Normalized emitted light spectrum of the CdSe/ZnS quantum dot samples.

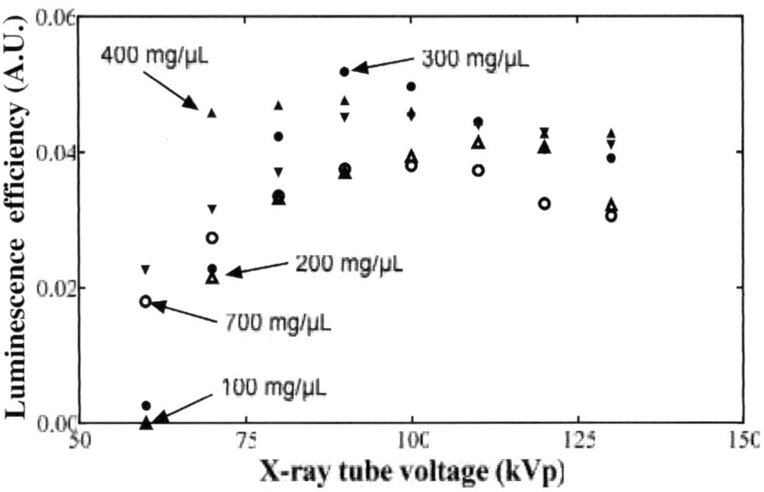

Figure 2. Luminescence efficiency variation with increasing concentration of the CdSe/ZnS quantum dots, in the range from 50 to 130 kVp.

4. Conclusions

In the present study preliminary results on the luminescence efficiency (LE) of toluene dissolved Cadmium Selenide/Zinc Sulfide (CdSe/ZnS, Sigma-Aldrich, Lumidot 694622) quantum dot samples (QDs), after exposure to X-rays are reported. The influence of the weight over volume (w/v) concentration of the samples in LE was investigated. The w/v concentration of the QDs varied between 7.1×10^{-5} mg/mL to 28.4×10^{-5} mg/mL. The distribution of the emitted light was symmetrical with a maximum at 590 nm. The X-ray LE, induced by the 28.4×10^{-5} mg/mL QDs found higher. The maximum efficiency was obtained at the 90 kVp for the QDs of the 21.3x10^{-5} mg/mL w/v concentration. In the high energy range (120-130 kVp) all concentration levels exhibited comparable X-ray induced LE. The luminescence properties of the investigated QDs appear promising for X-ray detection applications.

5. References

[1] Antonuk L E 2006 *Phys. Med. Biol.* **47** R31.
[2] Antonuk L El-Mohri Y and Wang Y 2000 *Med. Phys.* **27** 289.
[3] Doi K 2006 *Phys. Med. Biol.* **51** R5.
[4] Rossa W, Cody D, Hazle J 2006 *Med. Phys.* **33(6)** 1888.
[5] Gupta R, Grasruck M, Suess C, Bartling S, Schmidt B, Stierstorfer K, Popescu S, Brady T, Flohr T 2006 *Eur. Radiol.* **16** 1191.
[6] Yaffe M, Mainprize J, Jong R 2008 *Health Phys.* **95(5)** 599.
[7] van Eijk C 2002 *Phys. Med. Biol.* **45** R85.
[8] Michail C, David S, Liaparinos P, Valais I, Nikolopoulos D, Kalivas N, Toutountzis A, Sianoudis I, Cavouras D, Dimitropoulos N, Nomicos C, Kourkoutas K, Kandarakis I, Panayiotakis G 2007 *Nucl. Instrum. Meth. Phys. Res.* A **580**, 558.
[9] Michail C, Valais I, Seferis I, Kalyvas N, David S, Fountos G, Kandarakis I 2014 *Radiat. Meas.* **70** 59.
[10] Blasse G and Grabmaier B 1994 *Luminescent materials* (Berlin:Spinger).
[11] Michail C, Kalyvas N, Valais I, David S, Seferis I, Toutountzis A, Karabotsos A, Liaparinos P, Fountos G and Kandarakis I 2013 *J. Lumin.* **144** 45.
[12] Nagarkar V, Miller, S, Tipnis S, Lempicki A, Brecher C, Lingertat H 2004 *Nucl. Instr. Meth. B.* **213** 250.
[13] Michail C, Fountos G, David S, Valais I, Toutountzis A, Kalyvas N, Kandarakis I and Panayiotakis G 2009 *Meas. Sci. Technol.* **20** 104008.
[14] Michail C, Kalyvas N, Valais I, Fudos I, Fountos G, Dimitropoulos N, Koulouras G, Kandris D, Samarakou M, Kandarakis I 2014 *Biomed. Res. Int.* **2014** 634856.
[15] Kandarakis I and Cavouras D 2001 *Nucl. Instrum. Meth. Phys. Res. A* **460** 412.
[16] Seferis I, Michail C, Valais I, Fountos G, Kalyvas N, Stromatia F, Oikonomou G, Kandarakis I, Panayiotakis G 2013 *Nucl. Instrum. Meth. Phys. Res. A* **729** 307.
[17] Seferis I, Michail C, Valais I, Zeler J, Liaparinos P, Fountos G, Kalyvas N, David S, Stromatia F, Zych E, Kandarakis I, Panayiotakis G 2014 *J. Lumin.* **151** 229.
[18] Zych E, Meijerink A and Doneg C 2003 *J. Phys.: Condens. Matter* **15** 5145.
[19] Kim S, Park J, Kang S, Cha B, Cho S, Shin J, Son D, Nam S 2007 *Nucl. Instrum. Meth. Phys. Res. A* **576** 70.
[20] Konstantatos G, Clifford J, Levina L and Sargent E 2007 *Nat. Photonics* **1**, 531.
[21] Konstantatos G and Sargent E 2010 *Nat. Nanotechnol.* **5** 391.
[22] Rauch T, Böberl M, Tedde S, Fürst J, Kovalenko M, Hesser G, Lemmer U, Heiss W and Hayden O 2009 *Nat. Photonics* **3** 332.
[23] Eychmuller A 2000 *J. Phys. Chem. B* **104** 6514.
[24] Ma Y, Qi L, Ma J, Cheng H, Shen W 2003 *Langmuir* **19** 9079.
[25] Guo S, Konopny L, Popovitz-Biro R, Cohen H, Sirota M, Lifshitz E, Lahav M 2000 *Adv. Mater.* **12** 302.
[26] Wang C, Chen A, Chen I 2006 *Polym. Adv. Technol.* **17** 598.
[27] Michail C, Spyropoulou V, Fountos G, Kalyvas N, Valais I, Kandarakis I, Panayiotakis G 2011 *IEEE Trans. Nucl. Sci.* **58(1)** 314.
[28] Michail C, Valais I, Seferis I, Kalyvas N, Fountos G and Kandarakis I 2015 *Radiat. Meas.* **74** 39.
[29] Lumidot CdSe/ZnS 610, core-shell type quantum dots, specification sheet, 694614 Sigma-Aldrich, http://www.sigmaaldrich.com/catalog/product/aldrich/694614?lang=en&region=GR

G-autonomy of EEG recordings of psychotic patients undergoing the primitive expression form of dance therapy

E-C Ventouras[1], I Lardi[1], S Dimitriou[1], A Margariti[2,3], P Chondraki[2], I Kalatzis[1], N-T Economou[2], H Tsekou[2], T Paparrigopoulos[2] and P Y Ktonas[2]

[1] Department of Biomedical Engineering, Technological Educational Institution of Athens, Agiou Spyridonos Street, Athens, 12210, Greece

[2] Department of Psychiatry, 1st Psychiatric Clinic, Eginition Hospital, Medical School, University of Athens, 74 Vassilissis Sophias Avenue, Athens, 11528, Greece

[3] Department of Theater Studies, University of Peloponnese, 21 Vassileos Konstantinou Street, Nafplion, 21460, Greece

Abstract. Primitive expression (PE) is a form of dance therapy (DT) that involves an interaction of ethologically and socially based forms which are supplied for re-enactment. Brain connectivity has been measured in electroencephalographic (EEG) data of patients with schizophrenia undergoing PE DT, using the correlation coefficient and mutual information. These parameters do not measure the existence or absence of directionality in the connectivity. The present study investigates the use of the G-autonomy measure of EEG electrode voltages of the same group of schizophrenic patients. G-autonomy is a measure of the "autonomy" of a system. It indicates the degree by which prediction of the system's future evolution is enhanced by taking into account its own past states, in comparison to predictions based on past states of a set of external variables. In the present research, "own" past states refer to voltage values in the time series recorded at a specific electrode and "external" variables refer to the voltage values recorded at other electrodes. Indication is provided for an acute effect of early-stage PE DT expressed by the augmentation of G-autonomy in the delta rhythm and an acute effect of late-stage PE DT expressed by the reduction of G-autonomy in the theta and alpha rhythms.

1. Introduction

Dancing belongs to a time-honoured form of therapeutic experience and practice. Dance therapy (DT) is the psychotherapeutic use of movement and dance through which the individual participates creatively in a process that furthers his cognitive, emotional, physical and social integration [1,2]. In the present work a specific form of DT called Primitive Expression (PE) is investigated. In PE DT, with the use of percussion, roles are played and figures of myth are enacted providing opportunities related to expression of feelings, use of unfamiliar stances and behaviours, which might lead to therapeutic experiences [2].

DT, when applied to patients undergoing treatment is usually evaluated by assessing psychological and behavioural modifications that are potentially connected with the DT process. The scope of the studies and their assessment horizon can be expanded if the research protocol includes also neurophysiological changes. In [3] stress hormone levels were measured in fibromyalgia patients undergoing DT. In [1] hormonal and neurotransmitter levels were measured in depressed non-medicated adolescents. In [2] a PE DT-based protocol was applied to a small group of medicated psychiatric patients with psychotic, obsessive compulsive and depressive disorders. Electroencephalographic (EEG) recordings were acquired in a subset of the patients suffering from psychotic disorders. PE DT treatment led to quantifiable modifications in psychological state, behaviour and to brain physiology as expressed by EEG recordings. It was found that the patients experienced an increase in their happiness level and expressed a positive attitude to the PE DT process. For the group of patients that had EEG recordings, there was an increase in the alpha EEG frequency activity (8–12 Hz).

A further step in the investigation of neurophysiological parameters has been made in a recent study [4], as an expansion of the study initiated in [2]. Correlation coefficient (ρ), mutual information (MI) and a set of novel measures extracted from ρ and MI were computed from the EEG of schizophrenic patients that underwent PE DT. Those parameters were computed with the aim of providing insights about brain connectivity changes that could have resulted from the PE DT process. Indication was provided for an acute potentiation effect, at late-stage PE DT, on the inter-hemispheric connectivity in frontal areas, as well as for attenuation of the inter-hemispheric connectivity of left frontal and right central areas. In the transition from early to late-stage PE DT, indication was provided for potentiation of the intra-hemispheric connectivity of frontal and central areas, bilaterally.

ρ and MI are concerned with undirected functional connectivity, while, most likely, a more in-depth understanding of neural processes might require investigation of directed functional connectivity. In the framework of linear regression modelling, directed connectivity might be investigated using the notion of Granger causality or G-causality [5,6]. Causality between two processes X1 and X2, expressed through two time series, can be substantiated, for example, if the inclusion of past observations of X2 reduces the prediction error of X1 in a linear regression model of X1 and X2, as compared to a model which includes only previous observations of X1. It might then be conjectured that X2 "causes" X1. The G-causality analytical framework can be used to measure the G-autonomy of a process [7]. In this perspective, autonomy is not thought as a simple reflection of lack of connectivity, which might have been indicated by low values of ρ and MI. Autonomy is related to the self-determination or 'self-causation' of the process investigated and the G-autonomy is augmented when the prediction error of X1 is reduced by including its own past observations, given a set of external variables X2 . . . Xn. In the present work, the EEGs of the same set of patients used in [4] were investigated for modifications in the level of G-autonomy of electrode recordings.

2. Materials and Methods

2.1. Subjects, EEG recordings and EEG pre-processing

The subjects' group, the EEG recording procedure and EEG pre-processing have been described in detail in [4]. The EEG recording of 8 medicated schizophrenic patients were used. 12 PE DT sessions took place. An EEG recording session, in the awake resting state with eyes closed, with recording electrodes located at F3, F4, C3, C4, O1 and O2, lasting for about 3 min, took place before the 5th DT session (s_1), after the 5th DT session (s_2), as well as before (s_3) and after (s_4) the 11th DT session. In the following, by "session" it will be meant "EEG recording session". For each patient and each session, an EEG data segment lasting 32 sec was selected for further analysis. If there were electrode recordings with unacceptably high level of artifacts in the duration of the selected segment, those electrodes were excluded from further analysis. The recorded data were filtered, with low-frequency and high-frequency cut-offs set at 0.4 Hz and 3.5 Hz, respectively, for the delta EEG rhythm, at 4 Hz and 7 Hz for the theta rhythm and at 8 Hz and 12 Hz for the alpha rhythm.

2.2. G-autonomy computation

The G-autonomy (ga) value for subject i (i=1,...,8), for electrode j (j=F4,C4,O2,F3,C3,O1), for session s_k (k=1,...,4) and for EEG rhythm r_m (m=D,T,A, corresponding to the delta, theta and alpha rhythm data, respectively) is denoted as $ga_{i,j}^{s_k,r_m}$. ga values and related parameters were computed using the "Granger causal connectivity analysis" (GCCA) freely available toolbox [8].

The time separation between EEG recording sessions s_2 and s_1 and between EEG recording sessions s_4 and s_3 was much smaller compared to the time separation between s_3 and s_1 and between s_4 and s_2. Therefore, it could be tentatively conjectured that the difference values

$$\Delta(ga_{i,j}^{s_{2-1},r_m}) \equiv ga_{i,j}^{s_2,r_m} - ga_{i,j}^{s_1,r_m} , \ \forall \text{ i,j,m} \tag{1}$$

might reflect an "acute" effect on the G-autonomy value occurring at the ("middle-stage") 5th DT session, and that the difference values

$$\Delta(ga_{i,j}^{s_{4-3},r_m}) \equiv ga_{i,j}^{s_4,r_m} - ga_{i,j}^{s_3,r_m} , \ \forall \text{ i,j,m} \tag{2}$$

might reflect an "acute" effect on the G-autonomy value occurring at the ("late-stage") 11th DT session.

For example, if a specific value of $\Delta(ga_{i,j}^{s_{2-1},r_m})$ was positive, this might indicate that for that specific subject, electrode and EEG rhythm, the "autonomy" of the processes reflected in the voltage time series, as far as it might be related to G-autonomy values, increased "just" after the completion of the 5th DT session, compared to the situation "just" before the 5th DT session.

Furthermore, it could be tentatively hypothesized that the difference values

$$\Delta(ga_{i,j}^{s_{3-1},r_m}) \equiv ga_{i,j}^{s_3,r_m} - ga_{i,j}^{s_1,r_m} , \ \forall \text{ i,j,m} \tag{3}$$

might reflect a "long-term" effect on brain connectivity, that occurred between just before the 5th and just before the 11th DT sessions. The difference values

$$\Delta(ga_{i,j}^{s_{4-2},r_m}) \equiv ga_{i,j}^{s_4,r_m} - ga_{i,j}^{s_2,r_m} , \ \forall \text{ i,j,m} \tag{4}$$

might reflect a "long-term" effect on brain connectivity, that took place between just after the 5th and just before the 11th DT sessions, and an "acute" effect on brain connectivity resulting from the ("late-stage") 11th DT session.

The same kind of difference values were also computed for the mean values $\overline{ga^{s_k,r_m}}$ of sessions k =1,…,4, of the distributions of $ga_{i,j}^{s_k,r_m}$.

3. Results and Discussion

In Table 1 the mean values $\overline{ga^{s_k,r_m}}$, and the respective standard deviations, per EEG recording session k=1,…,4 and rhythm (D,T,A) are given. Paired t-tests revealed that statistically significant differences existed for the differences $\overline{ga^{s_2,r_D}} - \overline{ga^{s_1,r_D}}$ =1.5 ($p<0.001$), $\overline{ga^{s_4,r_T}} - \overline{ga^{s_3,r_T}}$ =-1.4 ($p<0.001$), $\overline{ga^{s_4,r_A}} - \overline{ga^{s_3,r_A}}$ =-1.8 ($p<0.001$) and $\overline{ga^{s_3,r_D}} - \overline{ga^{s_1,r_D}}$ =1.2 ($p=0.002$). When the autonomy values per separate electrode recordings were investigated, no statistically significant differentiations emerged between sessions or between electrodes, per rhythm.

According to the rationale exposed in the previous section, it can be conjectured that indications are provided for an acute effect of early-stage PE DT expressed by the augmentation of G-autonomy in the delta rhythm, an acute effect of late-stage PE DT expressed by the reduction of G-autonomy in the theta and alpha rhythms and a long-term effect of PE DT expressed by the augmentation of G-autonomy in the delta rhythm that occurred just before the 5th and just before the 11th DT sessions.

Table 1. Mean values and standard deviation (s.d.) of *ga* for the various EEG rhythms and EEG recording sessions (s_1 to s_4). Asterisks denote the existence of statistically significant differences between the mean values having the same number of asterisks.

	Delta rhythm		Theta rhythm		Alpha rhythm	
	mean	s.d.	mean	s.d.	mean	s.d.
s_1	25.8*	1.7	26.0	1.0	25.9	1.1
s_2	27.3*	1.2	26.2	1.0	25.8	1.3
s_3	27.0	0.9	25.8**	0.7	25.9***	0.6
s_4	25.8	3.6	24.4**	1.5	24.1***	1.7

The above results should be examined in conjunction with the results that were presented in [4], although attention should be placed on the fact that in the present study the "autonomy" of the processes, probably reflected by the voltage recordings at each electrode, is measured in connection to all other available electrodes concurrently, making direct comparisons with electrode pair and regional connectivity indications not possible.

The present study has limitations concerning the relatively few electrodes used, the small number of subjects, and the lack of a control cohort [4]. Unless these limitations are addressed, any modifications in the EEG-derived measures cannot be solely attributed to the DT protocol used in the present study.

4. Conclusion
The present work enriches the set of EEG-based measures used for quantifying effects of DT, by examining the G-autonomy measure. Both acute and long-term effects of DT could tentatively be indicated, concerning modifications of the autonomy level of the brain processes that are responsible for producing the recorded scalp EEGs.

Acknowledgments
We would like to thank the 1st Psychiatric Clinic of the University of Athens Medical School, Eginition Hospital, for providing the patients and overall encouragement for this study. In particular, we appreciate the contribution of Dr. E. Daskalopoulos and of Profs. E.Aggelopoulos, D.Dikeos, A.Pehlivanidis, N.Stefanis, and G.Vaslamatzis. The support of Profs. G.Papadimitriou and C.Soldatos is especially appreciated.

References
[1] Jeong Y J, Hong S C, Myeong S L, Park M C, Kim Y K and Suh C M 2005 *Int. J. Neurosci.* **115** 1711.
[2] Margariti A, Ktonas P, Hondraki P, Daskalopoulou E, Kyriakopoulos G, Economou N T, Tsekou H, Paparrigopoulos T, Barbousi V and Vaslamatzis G 2012 *Arts Psychother.* **39** 95
[3] Bojner-Horwitz E, Theorell T and Anderberg U M 2003 *Arts Psychother.* **30** 255
[4] Ventouras E-C, Margariti A, Chondraki P, Kalatzis I, Economou N-T, Tsekou H, Paparrigopoulos T, and Ktonas P 2015 *Cogn. Neurodyn.* **9** 231
[5] Granger C 1969 *Econometrica* **37** 424
[6] Ding M, Chen Y and Bressler S 2006 Granger causality: basic theory and application to neuroscience *Handbook of time series analysis* eds S Schelter S, M Winterhalder, J Timmer (Wienheim: Wiley) pp 438–460
[7] Seth A 2009 *Artif. Life* **16** 2
[8] Seth A K 2010 *J. Neurosci. Meth.* **186** 262

Software platform for managing the classification of error-related potentials of observers

P Asvestas[1], E-C Ventouras[1], S Kostopoulos[1], K Sidiropoulos[1], V Korfiatis[2], A Korda[2], A Uzunolglu[2], I Karanasiou[2,3], I Kalatzis[1] and G Matsopoulos[2]
[1]Department of Biomedical Engineering, Technological Educational Institution of Athens, Athens, Greece
[2]School of Electrical and Computer Engineering, National Technical University of Athens, Athens, Greece
[3]Institute of Communications and Computer Systems, Athens, Greece

E-mail: pasv@teiath.gr

Abstract. Human learning is partly based on observation. Electroencephalographic recordings of subjects who perform acts (actors) or observe actors (observers), contain a negative waveform in the Evoked Potentials (EPs) of the actors that commit errors and of observers who observe the error-committing actors. This waveform is called the Error-Related Negativity (ERN). Its detection has applications in the context of Brain-Computer Interfaces. The present work describes a software system developed for managing EPs of observers, with the aim of classifying them into observations of either correct or incorrect actions. It consists of an integrated platform for the storage, management, processing and classification of EPs recorded during error-observation experiments. The system was developed using C# and the following development tools and frameworks: MySQL, .NET Framework, Entity Framework and Emgu CV, for interfacing with the machine learning library of OpenCV. Up to six features can be computed per EP recording per electrode. The user can select among various feature selection algorithms and then proceed to train one of three types of classifiers: Artificial Neural Networks, Support Vector Machines, k-nearest neighbour. Next the classifier can be used for classifying any EP curve that has been inputted to the database.

1. Introduction

A significant part of human learning takes place through observation. Studies in observational learning suggest that the mechanisms through which observation contributes to learning are similar to the mechanisms contributing to learning through self-action. A significant role in such mechanisms is error processing, an aspect of performance monitoring necessary for detecting errors, as fast as possible, and for optimizing future response behavior. Previous research has shown that, when errors are committed, a negative deflection appears in electroencephalographic Evoked Potentials (EPs), the Error-Related Negativity (ERN) [1]. Such a deflection has also been recorded when persons observed other committing errors or observed the performance of an external agent, in the context of Brain-Computer Interfaces (BCIs).

There are several software platforms for acquisition and processing continuous and event-related EEG, such as the EEGLab [2]. These platforms can be used for general purpose tasks, but they do not provide tools for the classification of EPs. This paper presents the development of a dedicated, integrated software system for the classification of EPs of subjects who observe the execution of correct or incorrect responses of other people or agents. The system comprise data collection, feature extraction and classification sub-systems.

2. System Architecture

The proposed system is an integrated platform for the storage, management, processing and classification of the EPs recorded during error-observation experiments. The system is comprised of the data acquisition sub-system, the feature extraction sub-system and the classification sub-system.

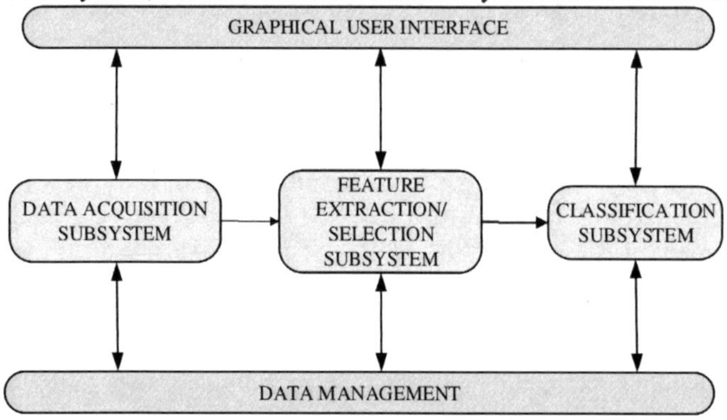

Figure 1. Architecture of the proposed system.

The data acquisition subsystem allows the acquisition of EPs from people who observe correct or incorrect actions of other subjects. The data are incorporated into a database along with various supporting data (demographics, date of recording, comments etc.).

The feature extraction/selection sub-system is responsible for the selection of features that are used for the description of EP recordings. Due to the relatively large amount of data obtained from EP recordings in modern electroencephalography (several tens of electrodes per person), appropriate signal processing techniques are available in the system for extraction and selection of a reduced set of quantitative indicators (features) to form a full description of the original data. Three methods for selecting features have been incorporated in the system: the Sequential Forward Search (SFS), the Sequential Backward Search (SBS) and the Sequential Floating Forward Search (SFFS) [3].

The classification subsystem categorizes vectors of features in one of the two categories of interest (observation of correct action and observation of incorrect actions). This subsystem can be regarded as implementing a function of many variables (classification function). The variables of this function are the components of each feature vector and the value of the function is the category of the feature vector. The classification system implements the k-Nearest Neighbor, the Support Vector Machine and the Artificial Neural Network techniques [3].

The system supports three different categories of users:

- simple users: psychologists or others without further technical knowledge who want to simply use the system to classify EP
- engineers: researchers with technical knowledge in signal processing and analysis, who have access to more advanced aspects of the system, such as feature selection and training of classifiers
- Administrators: users with full rights and access to each and every potential aspect of the system.

The system was developed using C# and the following development tools and frameworks:

- MySQL: database management system.
- .NET Framework: development platform for building graphical user interface and algorithms of signal processing and analysis using the C# programming language.
- Entity Framework: mapping tool of database tools to C# classes.
- Emgu CV: library machine learning algorithms.

3. Results

The initial screen of the application is shown in Figure 2. It contains two fields of entering the user name and password. A new user can be registered by selecting the link named Register. In this case, a registration form becomes visible (Figure 3). The user fills in the necessary information and submits the form.

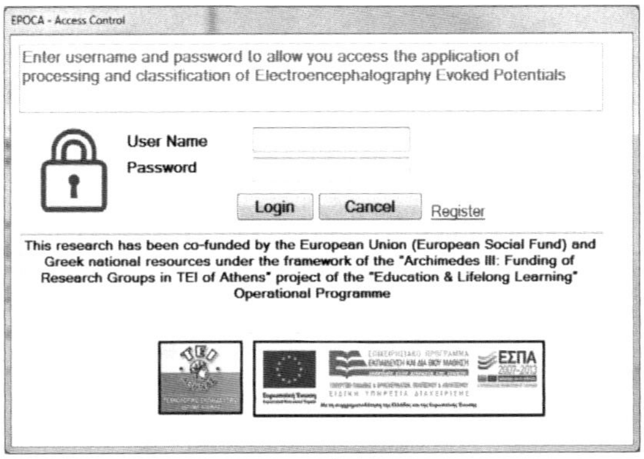

Figure 2. Access control form.

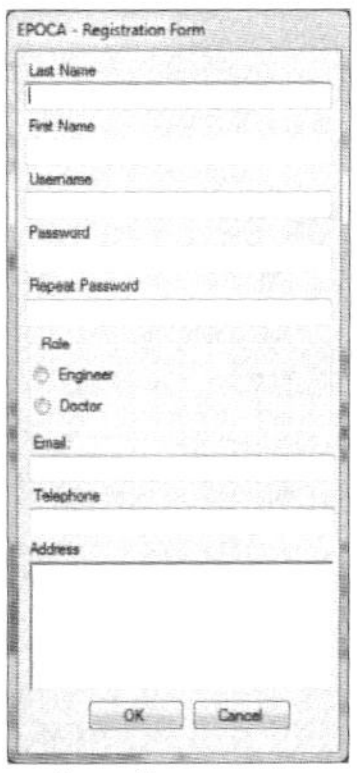

Figure 3. Registration form.

The main screen of the application is shown in Figure 4. It includes the following tabs: Data Management, Feature Extraction, Feature Selection, Classification, User Management and Settings. The Data Management tab displays the stored data in hierarchical order. The hierarchy consists of 3 levels. The first level includes subjects, the second level includes sessions and third level includes recordings (Figure 4). The next tab enables the extraction of various features (mean value, standard deviation etc.) from the available recordings (Figure 5). Optionally, smoothing filters (mean or median) and/or detrending can be applied before the feature extraction.

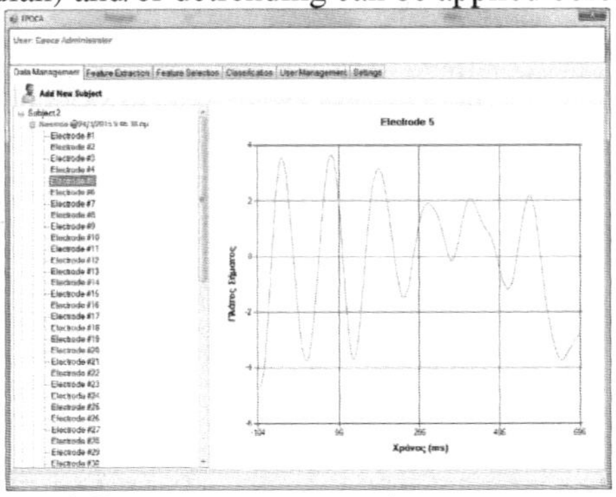

Figure 4. Data management tab.

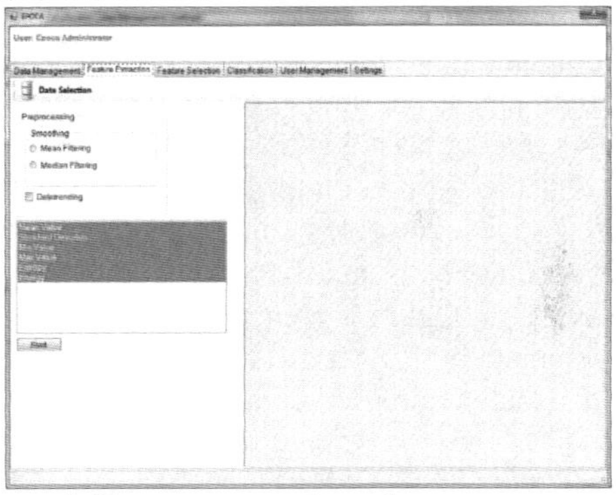

Figure 5. Feature extraction tab.

The third and fourth tabs include options for feature selection (Figure 6) and classification (Figure 7), respectively. The user management tab displays the registered users and provides reset password, update and delete capabilities (Figure 8). Finally, the last tab provides settings regarding the connection with the database server (Figure 9).

The evaluation of the performance of the classification subsystem has been presented in previous publications of the research team [4], [5].

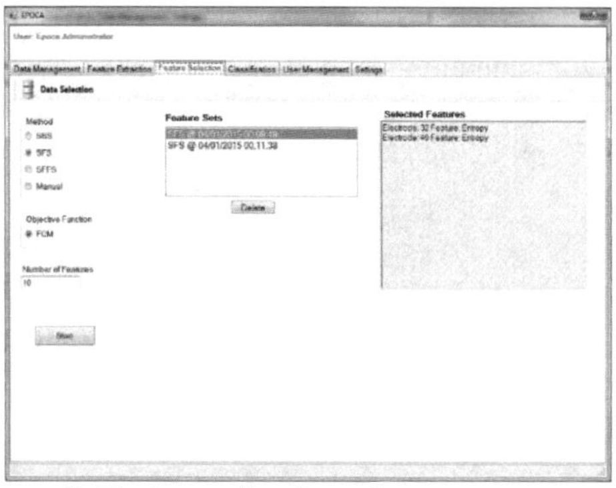

Figure 6. Feature selection tab.

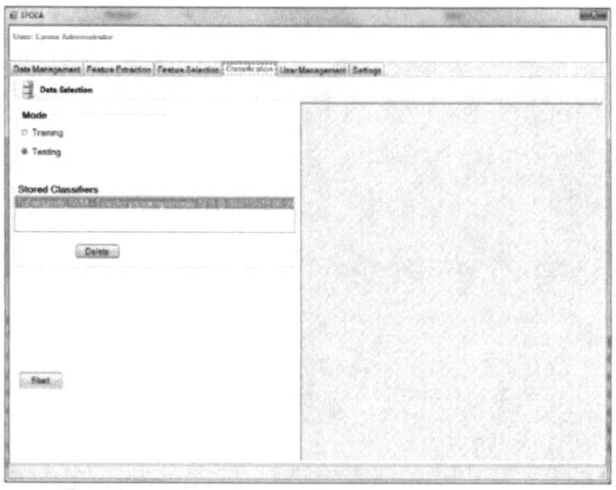

Figure 7.Classification tab.

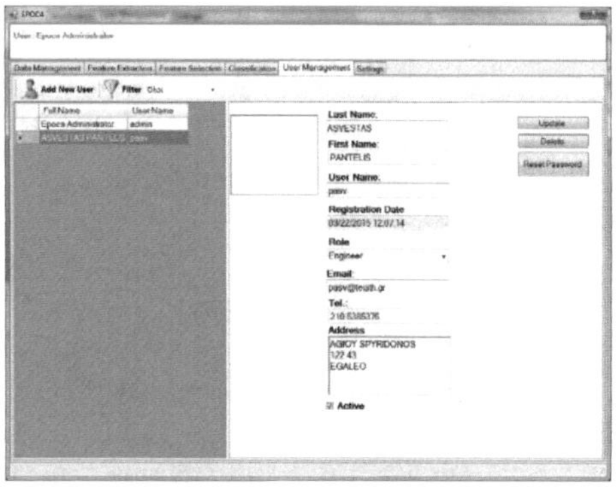

Figure 8. User management tab.

Figure 9. Settings tab.

Conclusion

Through the use of the proposed system, indications could be provided in real time concerning whether or not an observer, who oversees the actions of another person, perceives the existence of errors in the actions of the supervised person. Indications of critical parameters in display screens could also be the focus of observation, enabling the automatic evaluation of the performance of observers, which is significant for the selection of airplane pilots or operators of critical installations. The system might also contribute to the improvement of BCI performance, by enabling the automatic correction of wrong actions of the BCI, based on the EPs of the person using the BCI. In the future, it might also detect errors in joint actions, either between two persons, or between a person and a robot.

Acknowledgments

This research has been co-funded by the European Union (European Social Fund) and Greek national resources under the framework of the "Archimedes III: Funding of Research Groups in TEI of Athens" project of the "Education & Lifelong Learning" Operational Programme.

European Union
European Social Fund
OPERATIONAL PROGRAMME
EDUCATION AND LIFELONG LEARNING
investing in knowledge society
MINISTRY OF EDUCATION & RELIGIOUS AFFAIRS, CULTURE & SPORTS
MANAGING AUTHORITY
Co-financed by Greece and the European Union
NSRF 2007-2013
EUROPEAN SOCIAL FUND

References

[1] van Schie H, Mars R B, Coles M G H and Bekkering H 2004 *Nat. Neurosci.* **7** 549-54

[2] Delorme A and Makeig S (2004) *J. of Neurosci. Methods* **134** 9-21

[3] Theoridis S and Koutroumbas K 2006 *Pattern Recognition* (San Diego: Academic Press).

[4] Asvestas P, Ventouras E M, Matsopoulos G K and Karanasiou I 2014 *IFMBE Proc.* **43** 565-8

[5] Asvestas P, Ventouras E M, Karanasiou I and Matsopoulos G K 2013 *Comm. in Comp. and Inf. Sci.* **384** 40-9

Effect of common building materials in narrow shaped X-ray fields transmission

I Vlachos[1], X Tsantilas[2], G Fountos[3], H Delis[1], I Kandarakis[3] and G Panayiotakis[1]

[1]Department of Medical Physics, School of Medicine, University of Patras, Patras 26500, Greece
[2]Department of Radiology, Sismanoglio General Hospital of Athens, Athens 15126, Greece
[3]Department of Biomedical Engineering, Technological Educational Institution of Athens, Athens 12243, Greece

E-mail: panayiot@upatras.gr

Abstract. Diagnostic and interventional radiology, are an essential part of present day medical practice. Advances in X-ray imaging technology, together with developments in digital imaging have had a significant impact on the practice of radiology. This includes improvement in image quality, reduction in dose and a broader range of available applications resulting to better patient diagnosis and treatment. X-rays have the potential for damaging healthy cells and tissues, therefore all medical procedures employing X-ray equipment must be carefully managed. In all facilities and for all equipment types, procedures must be in place in order to ensure that exposures to patients, staff and the public are kept as low as reasonably achievable. Commonly used construction materials such as, ceramic tiles and plasterboards can provide a certain degree of protection against X-radiation. In this study, the secondary radiation transmission through common building materials is investigated, in the case of narrow shaped X-ray fields. Double plasterboard and double reinforced in thickness ceramic tile provided better radiation protection results.

1. Introduction

Diagnostic X-rays account for the major portion of man-made radiation exposure to the general population. Although individual doses associated with conventional radiography are usually low, examinations involving computed tomography and international radiography can be significantly higher. However, with well-designed, installed and maintained X-ray equipment, and through use of proper procedures by trained operators, unnecessary exposure to patients can be reduced significantly, with no degradation in the medical information derived. To the extent that patient exposure is reduced, there is, in general, a decrease in the exposure to the equipment operators and other health care personnel [1-3]. In some cases a judicious use of common building materials or a careful selection of location and orientation of the X-ray unit may eliminate the need for additional shielding without compromising the X-ray safety of the installation [4]. In this study the secondary radiation transmission through common building materials, such as single ceramic tile, double reinforced ceramic tile, single plasterboard and double plasterboard, for a narrow shaped X-ray field, is investigated [5,6]. As far as we know from current literature, this kind of combined study has not been performed previously. The presented results may be of use in panoramic radiography applications or other narrow field imaging techniques.

2. Materials and Methods

A conventional radiographic system (Philips Optimus 80) was used, installed at the Department of Radiology, Sismanoglio General Hospital of Athens, Greece. It was equipped with a three phase high voltage generator, a diagnostic X-ray tube with two focal spots, a tube voltage ranging from 40 to 150 kVp, a tube current ranging from 1 to 660 mA and an exposure time ranging from 0.001 to 16 s. The

Half Value Layer (HVL) was measured 2.1 mm Al at 70 kVp and the tube current was 10 mA. In order to simulate the human head, a cylindrical phantom from plexiglas, with a diameter of 16 cm and a height of 15 cm was used. The phantom was placed on the radiographic table and the distance between the phantom and the tube focus was 83 cm. The distance between the central point of phantom and the point of measuring secondary radiation was 50 cm. The measuring instruments were placed in a special radio protective apparatus and the materials studied were positioned at its input. The materials used as barriers, for the secondary X-ray transmission, were: (a) Single ceramic tile with thickness of 0.7 cm, (b) double reinforced ceramic tile with thickness of 1.8 cm, (c) single plasterboard [7] (or gypsumboard wall) with thickness of 1.3 cm and (d) double plasterboard with thickness of 2.6 cm. All thicknesses were measured with a Vernier caliper with reading error 0.05 mm. The X-ray field used was 16 cm x 2 cm, which is utilized for panoramic or other specific applications. An Amptek XR-100 CdTe spectrometer was used for measuring secondary X-rays energy distribution. The correction of the spectrometer with respect to the energy per X-ray bin was equal to (1/5.89) keV. In addition, the quantum efficiency response (QE) of the CdTe, obtained from the manufacturer data sheet, was equal to [8-10]:

$$Q(E) = 0.000002 * E^3 - 0.0001 * E^2 + 0.008 * E + 1.0362 \qquad (1)$$

where E is the X-ray energy. In X-ray spectrum measurements, the tube voltage was kept constant at 70 kVp and the tube load at 32 mAs. The ionization chamber used for measuring the secondary radiation was a calibrated 451P-DE-SI model of Fluke Biomedical. In this study, the irradiation time was kept at 2000 ms, so as to account for the response time of the instrument [11-13]. The secondary radiation dose rate (μSv/hr) was measured at a fixed location, with or without the barrier material studied. Before the measurements, a complete quality control program was performed on the radiographic unit to assess its reproducibility and accuracy.

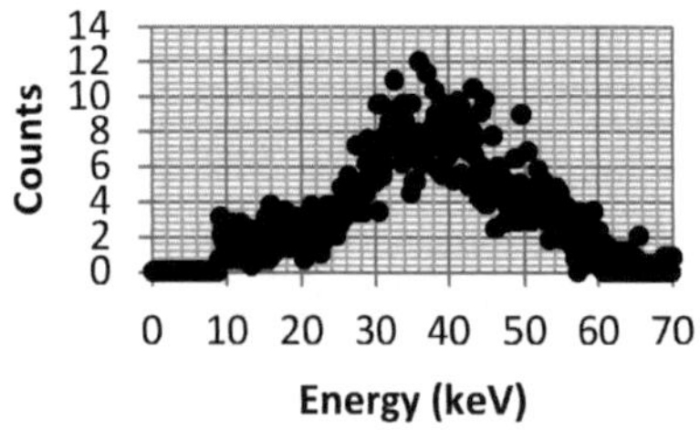

a)

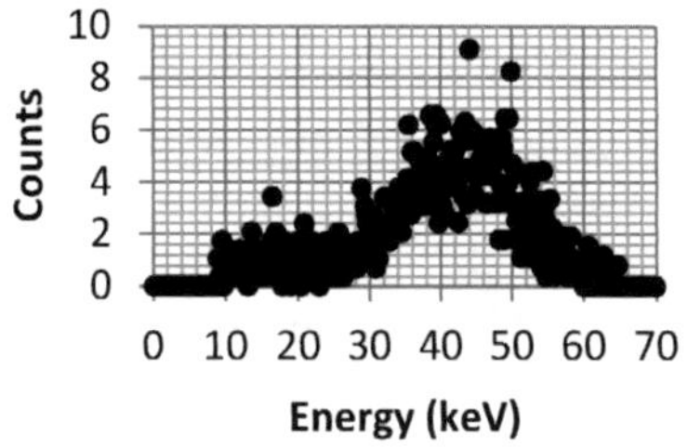

b)

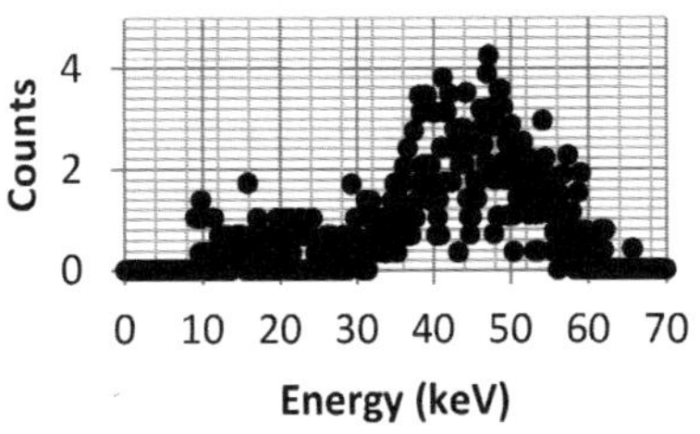

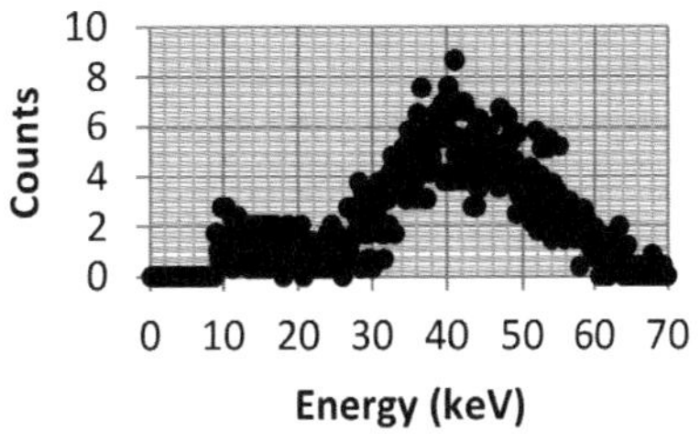

c) d)

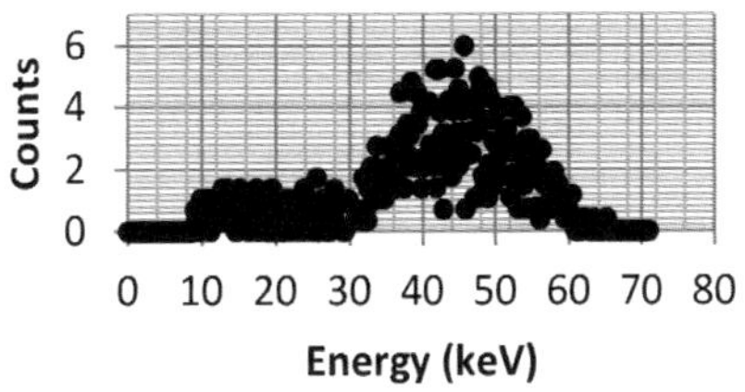

e)

Figure 1. Secondary X-ray energy distribution at 70 kVp, for: a) Without material, b) Single ceramic tile, c) Double reinforced ceramic tile, d) Single plasterboard, e) Double plasterboard.

3. Results and Discussion

The accuracy of the X-ray tube voltage was 1.3 %, and its reproducibility was 1.9 %. The measurements were performed with the PTW-Freiburg T43014 [14-15], Diavolt Universal kVp-Dose-time meter placed at a distance of 1.0 m from the focal spot on the upper surface of the cylindrical phantom.

In Figure 1, the secondary X-ray energy distributions at 70 kVp for all materials studied are presented. All materials slightly increase the average value of energy spectrum, as also presented in the third column at Table 1. Due to the directivity of the scatter radiation, the spectrometer alignment with respect to the incident X-rays and the hole diameter, the spectrometer was able to detect only a portion of the secondary X-rays, scattered at an angle of 90 degrees with respect to the primary X-rays.

Table 1. Secondary radiation dose rate, transmission and average energy for all materials studied.

	Secondary radiation dose rate (μSv/hr)	Transmission %	Average value of energy (keV)
Without material	500.0	-	37.3
Double plasterboard	150.0	30.0	41.5
Double reinforced in thickness ceramic tile	150.0	30.0	41.6
Single ceramic tile	200.4	40.1	39.6
Single plasterboard	300.0	60.0	40.0

In Table 1, the secondary radiation dose rate (μSv/hr), as well as the corresponding transmission, and the corresponding average spectrum energy are presented. Double plasterboard and double reinforced in thickness ceramic tile provided better radiation protection results. In both these cases the average secondary radiation energy value was higher, due to the beam hardening effect at these materials, as they have higher density. According to the above results, common building materials, such as the materials studied, may be of value for shielding purposes, especially in cases of low workload and spaces with small occupancy factor.

4. Conclusion

For narrow shaped X-ray fields, such as 16 cm x 2 cm, common building materials, such as single ceramic tile, double reinforced ceramic tile, single plasterboard and double plasterboard and the combination of them, may be of value for shielding purposes, especially in cases of low workload and spaces with small occupancy factor. From the above materials studied, the double plasterboard and the double reinforced in thickness ceramic tile provided better radiation protection with low cost, for narrow X-ray fields and for low energies.

References

[1] NCRP 2004 *Structural Shielding Design for Medical X-Ray Imaging Facilities - Report No. 147*

[2] Environmental Health Directorate Health Canada 1999 *Radiation Protection in Dentistry - Recommended Safety Procedures for the Use of Dental X-Ray Equipment - Safety code 30*

[3] Environmental Health Directorate Health Canada 1991 *Radiation Protection in Veterinary Medicine - Recommended Safety Procedures for Installation and Use of Veterinary X-Ray Equipment - Safety code 28*

[4] Petrantonaki M, Kappas C, Efstathopoulos E, Theodorakos Y and Panayiotakis G 1999 Calculating shielding requirements in diagnostic X-ray departments. *Br. J. Radiol.* **72** 179-185

[5] Edward R, Herbert W 1966 Experimental study of the photon energy spectrum of primary diagnostic X-rays *Phys. Med. Biol* **2** 225-238

[6] Archer B R, Thornby J I, Bushong S C 1983 Diagnostic X-ray Shielding Based in an Empirical Model of Photon Attenuation *Health Phys.* **5** 507-17

[7] See http://www.isolgomma.com/eng/

[8] Martini N, Koukou V, Michail C, Sotiropoulou P, Kalyvas N, Kandarakis I, Nikiforidis G, Fountos G 2015 Pencil beam spectral measurements of Ce, Ho, Yb, and Ba powders for potential use in medical applications *Journal of Spectroscopy* 563763

[9] Vlachos I, Tsantilas X, Kalyvas N, Delis H, Kandarakis I, Panayiotakis G 2015 Measuring scatter radiation in diagnostic X rays for radiation protection purposes *Rad. Prot. Dosim.* doi:10.1093/rpd/ncv093

[10] See http://www.amptek.com/

[11] Health Physics Society, http://www.hps.org/publicinformation/

[12] Economides S, Hourdakis C, Kalyvas N, Kalathaki M, Simantirakis G, Tritakis P, Manousaridis G, Vogiatzi S, Kipouros P, Boziari A, Kamenopoulou V 2007 Performance of medical radiographic X-ray systems in Greece for the time period 1998-2004 *Phys. Medica* **23** 107-114

[13] See http://www.flukebiomedical.com/

[14] See http://www.ptw.de/

[15] Tsalafoutas I 2006 Excessive leakage radiation measured on two mobile X-ray units due to the methodology used by the manufacturer to calculate and specify the required tube shielding *Br. J. Radiol.* **79** 162–164

TGF-β induced epithelial-mesenchymal transition modeling

P Xenitidis[1], I Seimenis[1], S Kakolyris[2], and A Adamopoulos[1]
[1] Medical Physics Lab, School of Medicine, Democritus University of Thrace, Alexandroupolis, Greece
[2] Department of Medical Oncology, School of Medicine, Democritus University of Thrace, Alexandroupolis, Greece

E-mail: pxenitid@med.duth.gr

Abstract. Epithelial cells may undergo a process called epithelial to mesenchymal transition (EMT). During EMT, cells lose their epithelial characteristics and acquire a migratory ability. Transforming growth factor-beta (TGF-β) signaling is considered to play an important role in EMT by regulating a set of genes through a gene regulatory network (GRN). This work aims at TGF-β induced EMT GRN modeling using publicly available experimental data (gene expression microarray data). The time-series network identification (TSNI) algorithm was used for inferring the EMT GRN. Receiver operating characteristic (ROC) and precision-recall (P-R) curves were constructed and the areas under them were used for evaluating the algorithm performance regarding network inference.

1. Introduction

The epithelial-mesenchymal transition (EMT) is a process that leads epithelial cells to lose their cell to cell adhesion and, thus, become migratory. Epithelial cadherin (E-cadherin) is a protein that is encoded by the CDH1 gene. Downregulation of E-cadherin decreases cellular adhesion in a tissue and is considered to play an important role in the EMT. There are several pathways that are involved in EMT such as TGF-β, Wnt, NF-kB and Notch [1][2].

In a cell, there are thousands of genes that are up or down regulated in order to produce or stop producing the appropriate proteins that are needed. Some proteins, called transcription factors, are only used to activate other genes. These transcription factors regulate gene expression and the whole system can be viewed as a network called gene regulatory network (GRN). The process of using experimental data to identify how genes interact within a regulatory network is called network identification, network inference or reverse engineering. High-throughput technologies such as DNA microarrays are widely used to measure simultaneously the expression levels of large numbers of genes. Thus, there have been many efforts for GRN inference from microarray data [3]. This work aims at identifying the GRN associated with one of the main pathways involved in EMT by exploiting publicly available microarray data.

2. Modeling Approaches

A GRN is usually modeled as a network that is composed of nodes that represent genes and edges that represent the interactions between the genes involved. The GRN inference is based on two basic elements: network topology and network parameters. The network topology refers to the process of finding the edges. The network parameters refer to the strength and the type of these interactions. The basic steps for GRN inference are shown in Figure 1 and include an initial stimulation (perturbation) and a set of measurements of the gene expression levels. Many mathematical models have been used for the GRN inference. Among them Boolean networks, differential equations, Bayesian networks, Petri nets, cellular automata, neural networks and Gaussian models [3]. Ordinary differential equations and specifically the TSNI algorithm proposed by Bansal [4] are employed in this study.

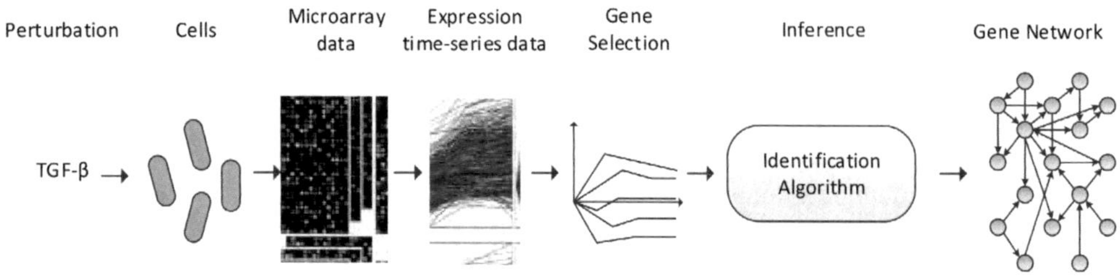

Figure 1. The basic steps for inferring a GRN.

3. Experimental Dataset and Pre-processing

The experimental dataset (GSE26858) that was used for the GRN inference was obtained from Gene Expression Omnibus (http://www.ncbi.nlm.nih.gov/geo/). It refers to the levels of expression at different time points due to TGF-β treatment of lung epithelial A549 cancer cells and contains four measurements at 0, 2, 12, and 24 h, respectively. At time t=0, the cells are treated with 2 ng/mL of TGF-b [5]. The data were interpolated using piecewise cubic interpolation. Principal component analysis (PCA) can be used in combination with TSNI algorithm for dimensionality reduction [4]. Therefore, a pilot study was performed in relation to the use of PCA on the TGF-β treatment dataset.

Firstly, PCA was applied on the original dataset of expression profiles. The relevant PCA scatterplot is shown in Figure 2(a). Then, genes demonstrating expression profiles with a variance lower than the 10th percentile were filtered out. Genes having very low absolute values (specifically lower than 10% of the whole data set) were also filtered out. The PCA scatterplot of the filtered expression profiles is shown in Figure 2(b).

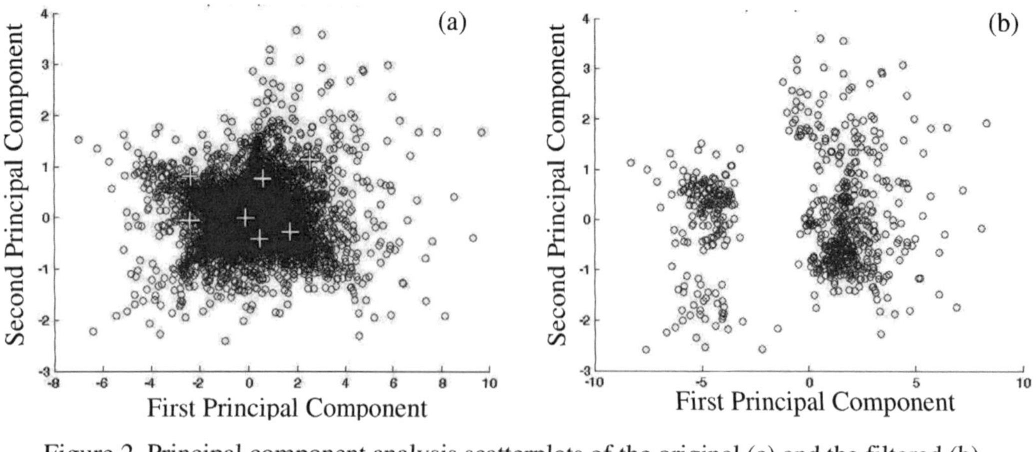

Figure 2. Principal component analysis scatterplots of the original (a) and the filtered (b) dataset. Marked dots in (a) correspond to genes known to be involved in EMT.

It is evident that filtering eliminates noisy gene expressions mainly located at the center of the scatterplot. Figure 2(b) also demonstrates four distinct gene clusters. If the expression profiles of the genes belonging to each cluster are plotted, then a similar pattern in each cluster is revealed. In Figure 2(a), the z-scores that belong to a group of genes that are widely accepted to be involved in the TGF-β induced EMT GRN have been superimposed on the original dataset. It is elicited that although PCA can identify strong expression patterns in a dataset, it lacks the power to reveal a group that is specifically related to the EMT process. It was decided, therefore, not to use the principle component analysis as a method for the assessment of the genes taking part in the inference process.

4. TSNI algorithm evaluation

The performance evaluation of an inference algorithm is typically based on the receiver operating characteristic (ROC) and the precision-recall (PR) curves [6]. ROC and PR curves, along with the areas under them (AUROC and AUPR, respectively) can be found in the supplementary material (Figures S1- S6 and Table S1) for all cases studied.

Initially, the TSNI algorithm performance was tested on artificial data. An artificial GRN containing 10 genes was generated. One gene (Gene 1) was only triggered and the TSNI algorithm was implemented to infer the artificial network. A good agreement was found between the generated and inferred connectivity matrices as far as Gene 1 coefficients are concerned. A poor match was found for the rest of the connectivity coefficients. Although the effect of the triggered gene on the rest of the genes in the GRN is well identified, the genes that were not triggered in the training dataset are not adequately identified. The same conclusion can be reached by simulating all gene expressions in the time domain.

To further explore the reasons behind the above mentioned poor match between generated and inferred data sets, a multi-gene triggering scenario was simulated. In this approach, all genes in the GRN were triggered at different time points (i.e., in a successive time order). Specifically, each gene was triggered with a pulse function type stimulation which allowed the triggered gene to evolve dynamically and provide it with adequate time to reach equilibrium. Afterwards, the next gene was triggered and so on. The time series data matching turned out excellent. A good matching was also found between the generated and inferred connectivity matrices.

This work, therefore, provides evidence that a process of GRN inference which is based on data produced by single-gene triggering may not able to capture the dynamics of the whole network. Using datasets produced by multi-gene stimulation, at separate time slots, greatly improves the inference performance of the implemented TSNI algorithm.

5. TGF-β induced EMT modeling

Since PCA did not reveal any group of genes that could be characterized EMT relevant, gene selection was based on previously published works [7][8]. For the model studied, seven genes were selected. The CDH1 gene has a dominant role in the specific model since it encodes the protein E-cadherin. Down-regulation of E-cadherin is considered a hallmark of EMT. The next gene selected was the Hmga2 gene which is induced by TGF-β through the Smad proteins. Hmga2 induces expression of Snail. Snail protein is an established regulator of EMT and is encoded by the Snai1 gene which was also taken into account. Snai2, Twist1, Zeb1 and Zeb2 genes were also included in the model.

The TSNI algorithm was implemented for the inference of TGF-β induced EMT GRN and the relevant connectivity matrix A was estimated. To evaluate the inferred system for TGF-β treatment, a simulation was performed in the time domain. Gene expression profiles are plotted in Figure 3, and it is clear that there is a very good agreement between simulated results and microarray measurements.

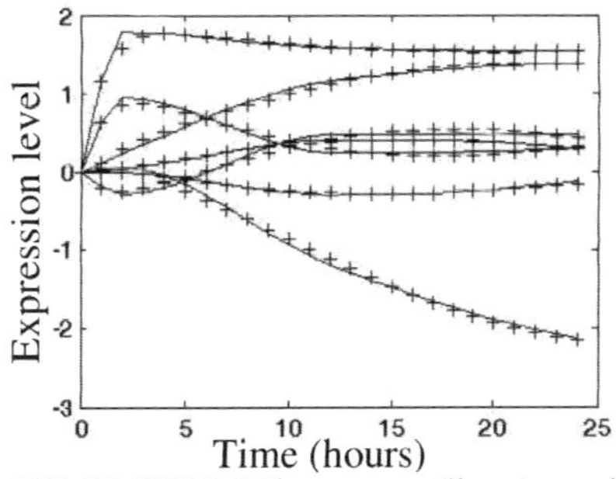

Figure 3. TGF-β induced EMT GRN. Microarray (lines) vs simulated (cross signs) data.

To evaluate the significance of the connectivity coefficients that were calculated by the TSNI algorithm, it was checked whether all coefficients have an impact on the time-series response of the inferred GRN. For this purpose, each coefficient a_{ij} of the inferred connectivity matrix A was varied and the mean square error (MSE) of each gene expression profile was calculated. All calculated MSEs were then added up to provide a total MSE index that describes how well the inferred system matches the original system in the time domain. After checking all the coefficient a_{ij} of the inferred connectivity matrix A, it was found that all coefficients were sensitive and, therefore, it is not convenient to set any coefficient equal to zero without significant total MSE increase. A representative MSE curve can be found in the supplementary material (Figure S7). The inferred system was also checked for stability by producing the relevant zero pole plot. All poles were found to lie within the left-half of the s-plane indicating that the system is stable.

6. Conclusions

This study focused on the modelling of the gene regulatory network that regulates the TGF-β induced EMT process. PCA proved inefficient in facilitating gene selection with regard to EMT. The performance of the TSNI algorithm chosen for GRN inference was tested in two cases. Single-gene perturbation and multi-gene perturbation at distinct time-points. It was shown that single-gene perturbation datasets lead to models that achieve to capture the dynamics of the system regarding the triggered gene but not its dynamics associated with the non-triggered genes. Therefore, we proposed a model with gene-specific triggering time to provide the algorithm with all the information needed for successfully identifying the connectivity matrix coefficients. It was demonstrated that the proposed approach significantly improves the entire modeling performance. The TGF-β perturbation is an intermediate scenario between single and multi-gene triggering. The model is expected to predict the GRN connectivity matrix with adequate accuracy for TGF-β treatment. It is postulated, however, that the model performance will be significantly deteriorated if it is tested with other types of stimulations.

7. Limitations

The fact that available microarray datasets on TGF-β treatment do not allow the performance of a multi-gene perturbation type training for the inferred GRN constitutes the main limitation of the current study. In addition, current work assumes that only 7 genes are closely related to the EMT process, an assumption which is not based on solid scientific data.

References

[1] Jing Y, Han Z, Zhang S, Liu Y and Wei L 2011 Epithelial-Mesenchymal Transition in tumor microenvironment *Cell & Bioscience* **1** 29

[2] De Herreros AG, Peiró S, Nassour M and Savagner P 2010 Snail family regulation and epithelial mesenchymal transitions in breast cancer progression *J. Mammary Gland Biol. Neoplasia* **15** 135–147.

[3] Hecker M, Lambeck S, Toepfer S, van Someren E and Guthke R 2009 Gene regulatory network inference: data integration in dynamic models-a review *Biosystems* **96** 86–103

[4] Bansal M, Della Gatta G and di Bernardo D 2006 Inference of gene regulatory networks and compound mode of action from time course gene expression profiles *Bioinformatics* **22** 815

[5] Zhang Y et al. 2011 High throughput determination of TGFβ1/SMAD3 targets in A549 lung epithelial cells *PLoS One* **6**

[6] Prill RJ et al. 2010 Towards a rigorous assessment of systems biology models: The DREAM3 challenges *PLoS One* **5**

[7] Tan E, Thuault S, Caja L, Carletti T, Heldin C and Moustakas A 2012 Regulation of Transcription Factor Twist Expression by the DNA Architectural Protein High Mobility Group A2 during Epithelial-to-Mesenchymal Transition *J. Biol. Chem.* **287** 7134–45

[8] Tran DD, Corsa CAS, Biswas H, Aft RL and Longmore GD 2011 Temporal and spatial cooperation of Snail1 and Twist1 during epithelial-mesenchymal transition predicts for human breast cancer recurrence *Mol. Cancer Res.* **9** 1644-57

Towards the estimation of the scattered energy spectra reaching the head of the medical staff during interventional radiology: A Monte Carlo simulation study

Zagorska A[1], Bliznakova K[2], Buchakliev Z[3]

[1] Department of Medical Physics and Biophysics, Medical University, 1 G. Sofiisky str, 1431 Sofia, Bulgaria
[2] Department of Electronic Engineering and Microelectronics, Technical University of Varna, Bulgaria
[3] SSDL-Sofia, National Centre of Radiobiology and Radiation Protection, 3 G. Sofiisky str, 1431 Sofia, Bulgaria

E-mail: zagorska.anna@gmail.com

Abstract. In 2012, the International Commission on Radiological Protection has recommended a reduction of the dose limits to the eye lens for occupational exposure. Recent studies showed that in interventional rooms is possible to reach these limits especially without using protective equipment. The aim of this study was to calculate the scattered energy spectra distribution at the level of the operator's head. For this purpose, an in-house developed Monte Carlo-based computer application was used to design computational phantoms (patient and operator), the acquisition geometry as well as to simulate the photon transport through the designed system. The initial spectra from 70 kV tube voltage and 8 different filtrations were calculated according to the IPEM Report 78. An experimental study was carried out to verify the results from the simulations. The calculated scattered radiation distributions were compared to the initial incident on the patient spectra. Results showed that there is no large difference between the effective energies of the scattered spectra registered in front of the operator's head obtained from simulations of all 8 incident spectra. The results from the experimental study agreed well to simulations as well.

1. Introduction

In recent years there is an increase practice of interventional radiology, which results in increasing doses at unshielded parts of the body of medical specialists. In 2012, the International Commission on Radiological Protection has recommended a reduction of the dose limits to the lens of the eyes for occupational exposure [1]. With respect to these new limits, investigations of different parameters affecting radiation protection to medical staff in interventional procedures are necessary. The overall goal of a project that is running in the Department of Medical Physics and Biophysics, Medical University, Sofia is to estimate the dose distribution within the lens of the eye of the operator's head. This paper presents the results from the first phase of this study with purpose to examine the impact of additional filtration to the scattered spectra around the X-ray unit, in particular at the level of the head of the medical staff at a position close to the X-ray unit.

2. Materials and methods

2.1. *Simulations.*

Figure 1a shows the computational phantoms designed for the purposes of this study. The 'patient' was chosen to be part of the Mathematical MIRD type hermaphrodite phantom, resembling a torso – elliptical cylinder, simulated from water, with the following dimensions: length of 70 cm, width of 40 cm and thickness of 20 cm [2]. Legs and patient's head were not included in the patient model, since their contribution to the scattered X-ray field is negligible in the irradiation conditions studied.

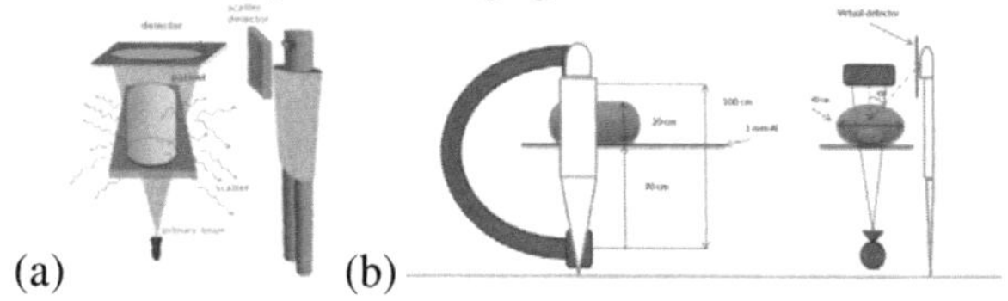

Figure 1. Simulation scenario: (a) schematic presentation of the phantoms used in the study and (b) simulated acquisition geometry.

The 'patient support coach' was modelled as a rectangular slab with dimensions of the rectangle equal to 100 cm and 60 cm, with thickness of 1 mm and composition set to aluminum (Al). The 'operator' was modelled as shown in figure 1a. The height of the examiner was consistent with the height of a standard sized human phantom – 176 cm [3]. The geometry used for the simulation of the interventional procedure is shown in figure 1b. Specifically, the focus-to-image detector distance was set to 100 cm, focus-to-table distance was set to 70 cm, while patient-to-scatter detector distance was equal to 20 cm. The X-ray source was modelled as a point source. Incident beam area on the detector was set to 400 mm^2 (200 mm x 200 mm). In addition to the designed system, a virtual detector with size of 500 mm x 500 mm, named scatter detector, was modelled and positioned in front of the operator's head in order to account for all scattered radiation reaching the operator's eye at a distance from the centre of the patient phantom to the centre of the scatter detector equal to 57 cm. The initial incident spectra were calculated for 70 kV tube voltage, 12° angle tungsten target, 3 mm and 4.5 mm Al inherent filtration with an additional copper filtration (Cu) (0.1 mm, 0.3 mm, 0.6 mm, 0.9 mm) according to IPEM 78 [4].

Further, to simulate the photon transport from the source through the designed system, as shown in figure 1a, the Monte Carlo module from the XRayImagingSimulator was used [5]. Only photons, reaching the scatter detector were recorded. The following detail information was concerned: photon energy, position and direction. This information will be used during the second phase of the project to evaluate the dose distribution within the eye lens of the operator. During simulation, the Kerma approximation was utilised, which is acceptable for photon energies below 200 keV, considered in this study [6].

2.2. *Laboratory test.*

To verify the results from the simulation studies, a laboratory test was carried out at the X-ray bench of the Secondary Standard Dosimetry Laboratory, Sofia. The 'patient' was simulated with a slab phantom with dimensions 300 mm x 300 mm x 150 mm [7]. An Al plate of 1 mm thickness was placed between the phantom and the X-ray tube to simulate the patient support coach, similarly to the geometry designed with the XRayImagingSimulator. Tube filtration and voltage were the same as these used in simulations. Ionizing chamber was situated at a position corresponding to the centre of virtual scatter detector to measure scattered air kerma (K_a). Electrometer Unidos10002 with Extradin ionizing chamber of 30 cm^3 volume, traceable to SSDL-Sofia, Bulgaria was used. Exposure and measurements were performed for all simulated filter combinations.

2.3. Data Analysis.

The overall average statistical uncertainty was calculated as $(\bar{E}) = \sqrt{\frac{1}{N}\left[\frac{1}{N}\sum_{i=1}^{N} e_i^2 - \bar{E}^2\right]}$, where $\bar{E}$ is the mean energy of the photons reaching the scatter detector, $e_i = \sum_j e_{ij}$, where e_{ij} denotes the energy of the jth photon of the ith history; the number of histories N varied between 7.10^6 and 4.10^7. Calculated overall average statistical uncertainties for all cases were less than 1%. All data processing and visualizations were realized in Matlab 8.1 [8]. First Half value layer (HVL_1) of the scattered spectra, as well as the air kerma (K_a), mean and effective energy were calculated. Data for $\frac{\mu_{en}}{\rho}$ of dry air were taken from NIST [9]. Calculated K_a from scattered spectra for all filter combinations was normalized to filter combination of Al3Cu0.1 and Al4.5Cu0.1, respectively.

3. Results and Discussion

Figure 2a,b show the calculated scattered spectra, obtained from simulation study with 3 mm Al, 4.5 mm Al and different Cu filters and registered at the scatter detector in comparison with the primary spectra.

(a) (b)
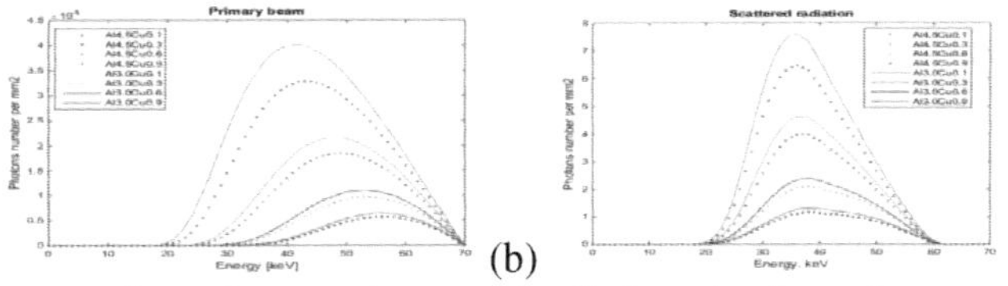

Figure 2. Comparison between primary beam and scattered radiation: a) primary beam spectra and b) simulated scattered spectra for 70 kV and different copper filters.

The shape of the spectrum depends on the voltage applied to the X-ray tube (kV), the angle, and the amount of inherent and added filtration in the X-ray beam. Aluminum filtration removes the low-energy end of the bremsstrahlung spectrum, which would otherwise be absorbed in the superficial tissues of the patient without contributing to the scattering process towards the operator's head. In practice an additional filter of copper is used to further harden the spectrum [10].

Calculated mean energy, HVL_1 and energy at maximum photons for primary beam and scattered radiation as well as effective energy of the simulated scattered spectra are summarised in table 1.

Table 1. Results from processing calculated data from primary and scattered spectra

Filtration	Primary beam			Scattered spectrum			
	HVL_1, mm Al	Mean energy, keV	Energy at maximum photons, keV	HVL_1, mm Al	Mean energy, keV	Energy at maximum photons, keV	Effective energy, keV
Al3Cu01	4.04	44.5	41.5	2.12	35.0	36	29
Al3Cu03	5.56	48.8	49.0	2.12	35.9	37	29
Al3Cu06	6.79	52.4	53.5	2.12	36.8	38	29
Al3Cu09	7.51	54.6	55.5	2.31	37.5	39	30
Al4.5Cu01	4.45	45.7	43.5	2.12	35.2	36	29
Al4.5Cu03	5.79	49.5	49.0	2.12	36.0	37	29
Al4.5Cu06	6.91	52.8	53.5	2.12	37.0	38	29
Al4.5Cu09	7.59	54.9	58.0	2.12	37.6	38	29

The results show that the mean energy of primary spectra varies between 45 keV and 55 keV with respect to the filtration while the respective mean energy of scattered spectra is in the range of 36 keV - 39 keV.

Calculated values of HVL_1 of different scattered spectra are similar. While the HVL_1 of the primary beam increases with additional filtration, the HVL_1 of the scattered spectra remains in the range of 2.1 mm – 2.3 mm Al. The effective energy of the scattered beam is calculated to be 29 keV-30 keV, not depending of the additional filtration.

In overall, good coincidence between results from laboratory tests and simulations is observed. Figure 3a,b summarise the results from the laboratory tests against simulations.

(a) (b)

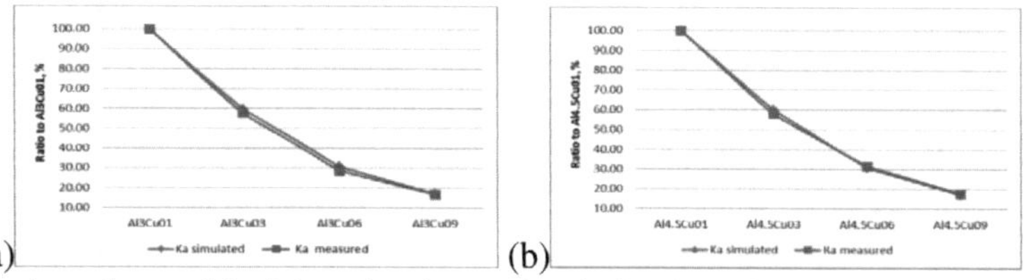

Figure 3. Normalised readouts of ionising chamber and calculated K_a for different filter combinations: (a) Al 3 mm, (b) Al 4.5 mm.

Results show that for filter combinations with 3 mm Al simulated and experimentally caluclated K_a diverge from 2 % to 7 %, while for filter combinations with 4.5 mm Al this discripency was between 3 % and 4 %.

4. Conclusion

This simulation study showed that for the studied tube voltage of 70 kVp and filter combinations, there is no large difference between the effective energy of the calculated scattered spectra registered in front of the operator's head. Simulation results agreed well with experimental ones. Presently, simulations are running with other tube potentials and different X-ray tube positions. The photon distributions registered in front of the operator head will be used to study the distribution of the absorbed dose in the operator's lenses.

5. Acknowledgements

This work was granted by Medical University – Sofia under Project 12D/2014, Contract 23D/2014.
The authors express sincere gratitude to clouds services by Oblak.bg for granted infrastructure to perform simulations.

6. References

[1] International Commission on Radiological Protection. Statement on tissue reactions approved by the commission on April 21. 2011. ICRP; 2011

[2] MIRD Pamphlet № 5, 1978 Society of nuclear medicine

[3] ICRP (International Commission on Radiological Protection), 2003. ICRP Publication 89: Basic Anatomical and Physiological Data For Use In Radiological Protection

[4] Cranley K et al, 1997 IPEM Report 78: Catalogue of Diagnostic X-ray Spectra and Other Data (cd-rom ed.) The Institute of Physics and Engineering in Medicine (IPEM), York, UK

[5] Bliznakova K, Kolitsi Z, Pallikarakis N, 2004, A Monte Carlo based software tool for radiotherapy investigations Nucl. Instr. Meth. B 222 pp. 445-461

[6] Siiskonen T, Tapiovaara M, et all, 2007, Monte Carlo simulations of occupational radiation doses in interventional radiology, The British Journal of Radiology, 80, 460-468

[7] International Organization for Standardization. X and gamma reference radiation for calibrating dosemeters and dose rate meters and for determining their response as a function of photon energy. Calibration of area and personal dosemeters and the measurement of their response as a function of energy and angle of incidence. 1999. ISO 4037-3

[8] Academic MATLAB ® license No. 980068

[9] NIST, 2012 ⟨http://physics.nist.gov/cgi-bin/Star/compos.pl⟩.

[10] International Commission on Radiation Units and Measurements, 2005. Patient dosimetry for X rays used in medical imaging. ICRU report 74. Bethesda, MD, U.S.A.

Limit of Detection in X-ray Diffraction Measurements of Tissue Equivalent Samples

Y Zheng [1], N Vassiljev [1], A Konstantinidis [2], J Griffiths [1,3] and R Speller [1]

[1]Department of Medical Physics and Biomedical Engineering, University College London, London WC1E 6BT, UK
[2]Diagnostic Radiology and Radiation Protection, Christie Medical Physics and Engineering, The Christie NHS Foundation Trust, Manchester M20 4BX, UK
[3]Institute of Biomedical Engineering, University College London, London WC1E 6BT, UK

E-mail: yi.zheng.11@ucl.ac.uk

Abstract. There is a suggestion of a new approach to mammography whereby following a conventional mammogram, the radiologist could interrogate suspicious regions using X-ray diffraction whilst the patient is still present and to establish the true extent of disease. A starting point for this work is to quantify the minimum detectable amount of breast cancer within a realistic thickness phantom. Perspex has a similar diffraction pattern to healthy breast tissue whilst water is similar to breast tumour, hence these two materials are used as tissue equivalent test objects for X-ray diffraction measurements. The preliminary results show linear agreement between the ratio of Perspex to water and the ratio of the diffraction peak intensities at 0.7 nm^{-1} and 1.5 nm^{-1}. The minimum detectable limit for a component of the two 'tissue' mix was found to be 4.1%. This suggests that X-ray diffraction can be used to quantify tissue like mixtures down to the 4.1% / 95.9% mix level and hence has a strong potential for delineating the extent of infiltration disease.

1. Introduction

There is a growing concern that current mammographic techniques do not satisfactorily indicate the limits of tumour involvement in infiltrating lesions [1]. This leads to insufficient tissue being removed at surgery, which can result in subsequent recurrences of breast cancer. X-ray diffraction (XRD) has been suggested to have a higher contrast than standard X-ray mammography, therefore making it a potential accurate non-invasive method of breast tumour mapping [2]. However, XRD requires a longer exposure time than mammography and results in a higher dose. One method to reduce the examination time is to reduce the number of XRD acquisitions, taking key point measurements rather than a blanket exposure. This can be achieved by using mammography guided XRD acquisitions. A starting point for this investigation is to establish the sensitivity of the technique to the fine infiltrating structures of disease that surround obvious lesions. So that it is able to validate the ability of XRD for tracing the extent of tumours in the boundaries between tumours and healthy tissues, which often

Journal of Physics: Conference Series **637** (2015) 012037 doi:10.1088/1742-6596/637/1/012037

contains very small amount of cancerous cells. Two experiments have been presented in this paper to investigate this sensitivity.

2. Materials and Methods

The X-ray source used in both experiments was a laboratory Molybdenum (Mo) target source with 30μm Mo filtration. This creates a pseudo-monoenergetic spectrum dominated by Mo Kα (17.4 keV) and Mo Kβ (19.6 keV) lines (see figure 1). Water and Perspex have similar XRD features to healthy breast tissue and cancerous tissues. The diffraction peak of Perspex at 0.6 nm^{-1} represents the peak of healthy tissues at 1.1 nm^{-1}, while the peak of water at 1.6 nm^{-1} corresponds to the same peak as cancerous tissues. Therefore water and Perspex were used to form the test objects for these experiments.

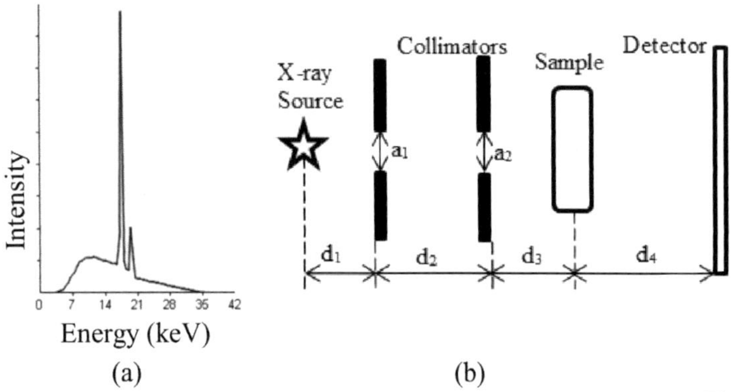

Figure 1. Experimental setup: (a) Energy spectrum of the X-ray source after the filtration; (b) schematic diagram of the geometry of the experimental setup

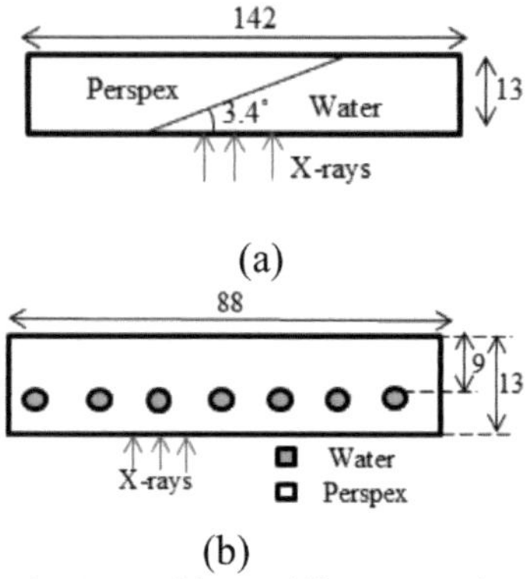

Figure 2. Test objects (distances given in units of mm): (a) the test object for experiment one; (b) the test object for experiment two

2.1. Experiment One

The aim of this experiment was to investigate the relationship between the XRD signal and the proportions of water and Perspex in the test object. The experimental setup is shown in figure 1. Distance d_1, d_2, d_3 and d_4 are 50 mm, 17 mm, 10.3 mm and 83.7 mm, respectively. The collimators are 3 mm thick Lead (Pb) blocks each with a circular hole of diameter (a_1 and a_2) 2 mm. The test object shown in figure 2 (a) was used in this experiment. A series of X-ray diffraction images corresponding to different proportion of water and Perspex were recorded.

2.2. Experiment Two

The second experiment was designed to investigate the minimum amount of water that can be detected by the XRD system. In other words, the aim of this experiment was to find the lower limit of detection of water. A second phantom (shown in figure 2 (b)) was designed for this purpose. It contained several holes (of diameter from 0.85 mm to 1.6 mm) filled with water within a Perspex block.

Furthermore, in this experiment, the distance between the different components in the experimental setup was altered to increase the intensity of the X-ray beam. In this experiment, d_1, d_2, d_3 and d_4 were 45 mm, 11 mm, 3.3 mm and 90.5 mm, respectively, while a_1 was 1.0 mm and a_2 was 1.2 mm.

2.3. Pre-processing

The background signal (an image without the presence of test object) was subtracted from XRD signal. Following this, pixels equidistant from the centre of the XRD pattern were averaged to transform the XRD signal from 2D data to a 1D profile.

3. Results and Discussions

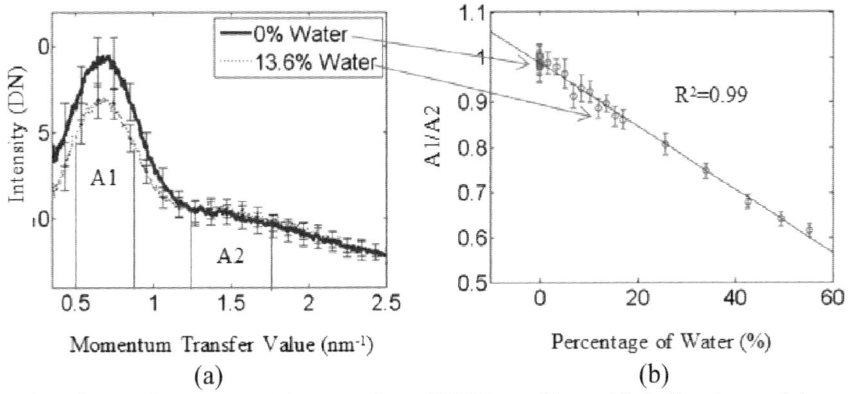

Figure 3. Results of experiment one: (a) examples of XRD profiles with indications of the positions of peak area A1 and A2. DN represents 'Digital Number'. (b) The ratios of peak areas (A1/A2) versus the percentage of water

3.1. Results from experiment one

The results from experiment one are indicated in figure 3. Figure 3(a) demonstrates two examples of XRD profiles for experiment one. Each curve corresponds to a data point in figure 3(b) indicated by the arrows. The error bars in figure 3(a) are the standard deviation of 10 measurements. The vertical axis in figure 3(b) represents the normalized ratio of the area under the diffraction peaks at 0.7 nm^{-1} and 1.5 nm^{-1} (A1/A2). The error bars in figure 3(b) are the propagation error of the standard deviation of 10 measurements. The propagation error in figure 3(b) can be calculated by:

$$\sigma_R = ((\sigma_{A1}/A1)^2 + (\sigma_{A2}/A2)^2)^{1/2} R$$

where R equals to A1/A2, σ_{A1} and σ_{A2} are the errors of peak area A1 and peak area A2, respectively. As can be seen in figure 3(b), there was a linear response between the ratio of the peak areas and the percentage of water. The gradient indicates the sensitivity of the ratio of peak areas A1/A2 to the presence of water.

3.2. Results from experiment two

The diffraction profiles of Perspex and the mixture of Perspex and water are shown in figure 4. The lines and error bars in figure 4(a) and (b) are the mean value and standard deviation of 12 measurements, respectively. From figures 4(a) and (b), we can see the difference between the XRD profiles of 0% water and 4 % water are very subtle, while the difference between Perspex and 11% of water is significant (outside the error bars).

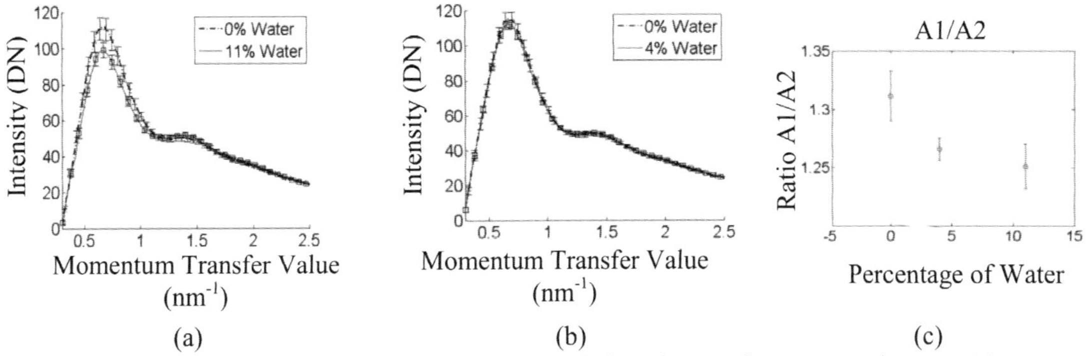

Figure 4. Comparison of XRD profiles of Perspex and the mixture of Perspex and water: (a) Perspex and 11% water; (b) Perspex and 4.1% water (c) the ratio of A1/A2 versus percentage of water in the test object

The statistics of the results can be improved by integrating the peak areas around 0.7 nm^{-1} and 1.5 nm^{-1}. The results are shown in figure 4(c). The error bars are the propagation error based on the standard deviation of 12 measurements. By calculating the ratio of the two peak areas, the difference between 0% water and 4.1%/4.6% water is greater than the error bar.

3.3. Results from transmission images
Although there were differences shown in XRD profiles when changing the proportion of water, the transmission image of the corresponding phantom did not show a significant change (as shown in figure 5). Notice that for the same position in figure 5(a), the effective percentage of water is different in the transmission image and the XRD images. This is because the size of the X-ray beam in the XRD experiment is larger than the diameters of the holes filled with water. The contrast at the position P (in figure 5(a)) is 0.029 ± 0.001 in the transmission image, while it is 0.048 ± 0.001 when using A1/A2 in the XRD profiles. Nevertheless, the difference of the linear attenuation coefficients between adipose and glandular breast tissue is bigger than that between Perspex and water. Therefore, the contrast of the transmission image in this experiment between different concentrations should be higher when using breast tissue.

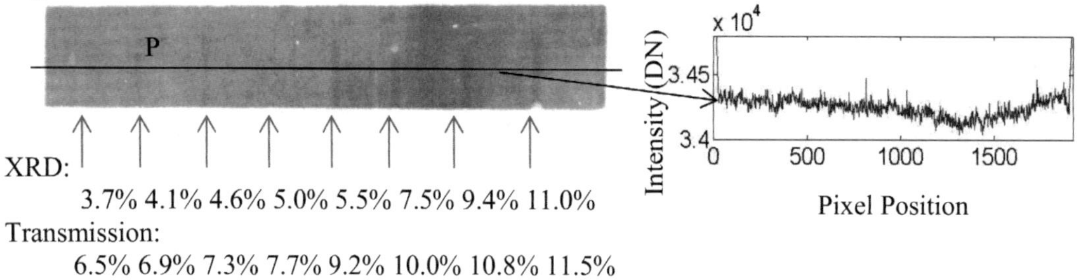

(a) (b)

Figure 5. (a) Transmission image of the test object used in experiment two; (b) Profile plot of a line indicated in (a)

4. Conclusions and Future Work
This experiment has shown that the lower limit of detection of water (representing breast tumour) in a mixture of water and Perspex (representing healthy breast tissues) is 4.1%. This indicates that XRD has a much better performance than transmission imaging, and thus suggests that the XRD system can provide much better contrast between tumour tissue and healthy tissue than mammography. However, in order to fully understand the lower detection limit of breast cancer, real breast cancer and healthy tissue will need to be investigated. Furthermore, a more realistic phantom with a thickness similar to breast (such as 50 mm) should also be investigated.

References
[1] Poletti M E, Gonçalves O D and Mazzaro I 2004 Measurements of X-ray scatter signatures for some tissue-equivalent materials *Nuclear Instruments and Methods in Physics Research Section B: Beam Interactions with Materials and Atoms* **213** 595-8
[2] Griffiths J A, Royle G J, Hanby A M, Horrocks J A, Bohndiek S E and Speller R D 2007 Correlation of energy dispersive diffraction signatures and microCT of small breast tissue samples with pathological analysis *Physics in medicine and biology* **52** 6151-64

Computational study of the optimum gradient magnetic field for the navigation of spherical particles into targeted areas

N K Lampropoulos[1], E G Karvelas[1,2], D I Papadimitriou[1] and I E Sarris[1]

[1]Department of Energy Technology, Technological & Educational Institute of Athens, Agiou Spiridonos 17, 12210 Athens, Greece.
[2]Department of Civil Engineering, University of Thessaly, Pedion Areos, Gr - 38334 Volos, Greece.

E-mail: Karvelas@uth.gr

Abstract. Spherical magnetic nanoparticles coated with drugs are navigated to targeted areas, for the treatment of cancer. The particles are navigated by magnetic field gradients that can be produced by an MRI device. In the present work, a computational study for the estimation of the time evolution of the gradient magnetic field is presented in order to ensure the optimum driving of the particles into the targeted area. For this purpose, the present method takes under consideration all the forces acting on the particles that make them move. The method is based on an iteration algorithm that intents to minimize the deviation of the particles from a desired trajectory. In this way, the gradient magnetic field is temporarily adjusted in a suitable way so that the particles' distances from the trajectory are decreased. Using the above mentioned method, it is clear that with the increase of the optimization parameters, i.e the modification of the gradient magnetic field, the particles are moved closer to the desired trajectory. Moreover, it is found that the present numerical model can navigate particles into the desired trajectory with efficiency above 90%.

1. Introduction

Drug delivery by magnetic nanoparticles in cancer tumors through blood vessels is considered. The benefits of using this method are numerous. Firstly, the quantity of the drug required to reach therapeutic levels is being reduced. Moreover, the drug concentration at targeted sites is increased. For the magnetic navigation of the drug, the use of magnetic nanoparticles is necessary. The anti-tumor agents are loaded on the surface of the magnetic nanoparticles. The magnetic field that is required in order to navigate the particles inside the human arteries is produced by a Magnetic Resonance Imaging (MRI) system.

The major factors, which influence the efficiency of the usage of magnetic nanoparticles for biomedical applications in the magnetic driving, are the size and the magnetization of the biocompatible nanoparticles. Many of these particles are being trapped in the liver and spleen and are excreted via the kidneys. Particles with small size imply a small magnetic response, thus the use of huge magnets are imperative [1]. Using paramagnetic nanoparticles, their magnetic response is maximized, as they are formed into chains under the influence of a steady magnetic field. The size of aggregates is very important and it depends on different parameters [2]. After the navigation into the infected area by gradient magnetic fields, the nanoparticles lose their magnetization in the absence of the magnetic field. The nanoparticles, which do not reach the tumor, continue to circulate into the human blood and are being opsonized and cleared by the macrophages.

Experimental data indicate the maximum value of the gradient magnetic field for steering the nanoparticles in the desired direction [3]. Due to the complex geometry of the arteries, the gradient magnetic field should be changed continuously.

2. Numerical Model

For the propulsion model of the particles seven major forces are considered; i.e. the magnetic force from MRIs Main Magnet static field, as well as the Magnetic field gradient force from the special Propulsion Gradient Coils. The static field caters for the aggregation of nanoparticles, while the magnetic gradient navigates the agglomerations. Moreover, two kind of forces, contact and drag forces are also considered. The contact forces are firstly developed between the aggregated nanoparticles and secondly between nanoparticles and walls. The Stokes' drag force for each particle is considered, while only spherical particles are considered in this study. In addition, forces due to gravity and buoyancy, are included. The numerical model for all the above mentioned forces is given in [4]. Finally, attractive forces, such as the Van der Walls forces, are included in the present model [5].

3. Numerical Method

The OpenFoam platform [6] was used for the calculation of the flow field and the uncoupled equations of particles motion. The simulation process goes as follows; initially, the fluid flow is determined using the incompressible Navier-Stokes equations and the PISO method. Upon finding the flow field (pressure and velocity), the motion of particles is evaluated by the Lagrangian method. The equations are solved in time by the Euler time marching method. The stability of the algorithm is guaranteed through a time step of the order of 10^{-6} s. The method is based on an iteration algorithm that intent to minimize the deviation of the particles from a desired trajectory, which starts from the middle of the inlet and passes through the middle of the desired tube. The trajectory is predefined in the computational platform by a 10degree polynomial. In a two-dimensional simulation, the gradient magnetic field is temporarily adjusted perpendicular to the desired trajectory. This happens in a suitable way such that the particles' distance from the trajectory to be decreased. To verify the optimal gradient magnetic field a Covariance Matrix Adaptation Evolution Strategy (CMAES) was used in order to drive the particles into the desired area [7].

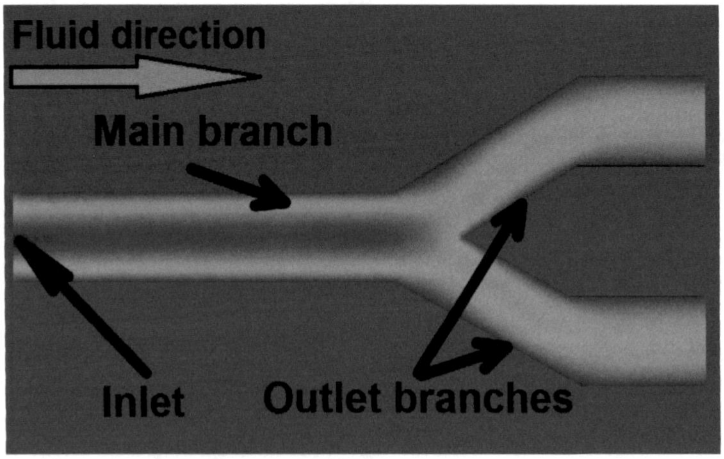

Figure 1. Y-shaped geometry.

4. Results

For the evaluation of the potentials of this numerical model, a series of simulations with different numbers of optimization parameters for the magnetic field were performed. For this reason, the combined flow of particles and water solution in a Y-shaped geometry (2D) was simulated, as shown in Figure 1. The inlet of the fluid is in the left of the geometry and there are two outlet branches in the right side. Between the two outlet branches is selected an angle of 60° degrees. Both diameters of the main and outlet branches are kept constant at 2.5 mm. The overall length of the simulated geometry was 0.036 m. The velocity of the fluid was 12.2 mm/s in the main branch of the geometry and 6.1 mm/s in the outlet branches [3]. One million particles were simulated. Each of the simulated particles has hydrodynamic diameter equal to 1 um and simulates aggregations of nanoparticles. The simulation was under a uniform transverse magnetic field of magnitude $B_0 = 1$ T.

Initially, the CMAES optimization strategy algorithm provides the OpenFOAM program with random values of gradient magnetic field. In the end of each simulation, the computational platform checks the distance between the particles and the desired trajectory. If all particles are in the desired trajectory, the simulation ends. If not, the optimization platform provides new values of gradient magnetic field in order to eliminate the distance of particles from the desired trajectory. In this way, the appropriate values of the gradient magnetic field are found for the particles' navigation into the targeted areas. Lower or higher values of the gradient magnetic field are rejected from the computational method, because in the first case the nanoparticles are not navigated into the desired area and in the second case, they are attached to the vessels' walls.

Table 1. Simulation Values

Time Step (s)	Color	mT/m
Initial	Red	0
0.2	Blue	-499.9
0.4	Pink	500
0.6	Cyan	152.9
0.8	Yellow	482.9
1	Orange	423.9
1.2	Grey	315
1.4	Black	9.8

From our calculations, it is clear that the present model can simulate the motion of the particles in a realistic manner and estimate the gradient magnetic field, in order to navigate the simulated particles into the desired trajectory, as it is depicted in Table 1. This model can drive the particles in prescribed areas, with efficiency more than 90% in the number of particles. On the other hand, a small amount of particles is stuck to the walls and remains there for the rest of the simulation, as it is shown in Figure 2.

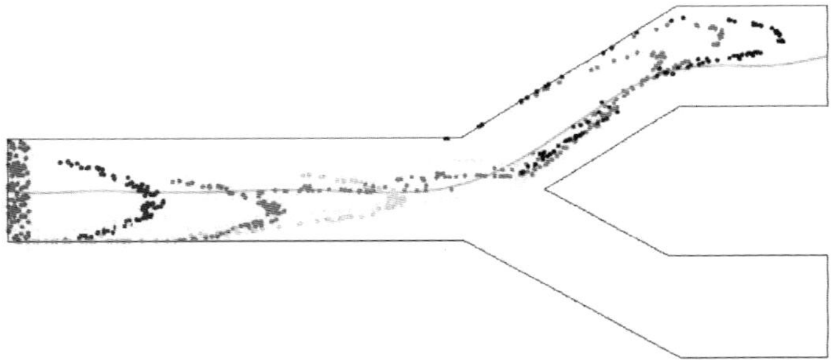

Figure 2. Desired trajectory (green line) and positions of each particle in each time step.

The same simulation as before was re-performed with more changes of the gradient magnetic field. Each one of these changes is considered as one optimization parameter. In the present study, the optimization parameters were increased from 8 to 18, i.e the corresponding changes of the gradient magnetic field. The particles' distance from the desired trajectory was measured just before the particles pass the outlet of the geometry.

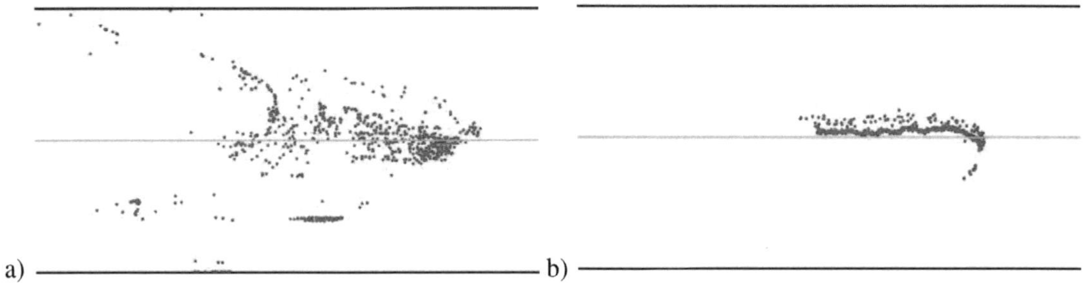

a) b)

Figure 3. Distribution of particles for: a) a simulation with 8 optimization parameters, b) a simulation with 18 optimization parameters. The black lines indicate the boundaries and the green line the desired trajectory.

The particles are getting closer to the desired trajectory, while the optimization parameters are increasing, as is showed in figure 3a,b. This comparison shows that the numerical method can drive successfully the particles into prespecified areas.

5. Discussion

In this work, a computational study of the optimum gradient magnetic field for the navigation of spherical particles into targeted areas is presented. The computational method can simulate an MRI system working under experimental conditions and drive particles into the desired trajectory with efficiency more than 90% in the number of particles. The results indicate that the more optimization parameters used, the smaller the distance from the desired trajectory is. Due to the fact that the particles' distance is minimized, we conclude that the present numerical model can drive the particles into specified areas. In order to eliminate particles' distances, more optimization parameters are needed. Also, it is known that the existed MRIs' can produce small changes in the gradient magnetic field at each instant. The computational method can simulate the gradient magnetic field of an MRI, when many optimization parameters are used. Thus, this requires more computational time.

Acknowledgments
The work is funded by the NANOTHER program (Magnetic Nanoparticles for targeted MRI Therapy) through the Operational Program COOPERATION 2011 of GSRT, Greece. Discussions with Dr Klinakis from BRFAA, Greece, Prof. Zergioti from NTUA, Greece and the people from Future Intelligence Ltd. are also acknowledged.

References

[1] Q. Pankhurst et al. 2003 *Physics D : Applied Physics* vol. **36** 166-181
[2] N.K. Lampropoulos, E.G. Karvelas and I.E. Sarris Computational study of particles interaction distance under the influence of steady magnetic field, Advances in Systems Science and Applications, in press 2015
[3] J-B Mathieu and S. Martel 2010 *Magn. Reson. Med.*, Vol. **63** 1336-45
[4] N.K. Lampropoulos, E.G. Karvelas and I.E. Sarris 2014 Computational modeling of an MRI guided drug delivery system based on magnetic nanoparticle aggregations for the navigation of paramagnetic nanocapsules, 11th World Congress on Computational Mechanics (WCCMXI) (Barcelona)
[5] P. Vartholomeos and C. Mavroidis 2012 *IEEE transactions on biomedical engineering*, vol. **59**. 3028 - 38
[6] H. G. Weller, G. Tabor, H. Jasak, C. Fureby, 1998, *Computers in physics*, Vol. **12**, No. 6.
[7] Hansen, N. 2006 *Studies in Fuzziness and Soft Computing* Vol. **192** 1769

An evaluation of the impact of digital imaging on radiographic practice and patient doses

J Horrocks and K Violaki

Clinical Physics, Barts Health NHS Trust, London E1 1BB

E-mail: julie.horrocks@bartshealth.nhs.uk

Abstract. Direct digital imaging technology was implemented in all areas in general and mobile radiology at Barts and the Royal London Hospitals in 2012. Evidence from recent radiation incident investigations indicates optimum exposure factors are not consistently selected, with the greater dynamic range of the digital detectors allowing sub-optimal practice. To investigate further patient dose data were extracted from the Radiology Information System for adult chest X-ray examinations in 2014, covering over 50,000 studies in the Trust. Chest X-ray examinations were selected as they are low dose but frequent examinations. The patient dose data were evaluated taking into account X-ray system type and detector performance measurements, and individual cases studies were used to highlight where practice can be improved.

1. Introduction

The use of electronic X-ray examination records contained in Radiology Information Systems (RIS) to audit patients doses and set Diagnostic Reference Levels (DRLs) has the advantage of allowing large numbers of examinations to be audited, but is prone to manual data entry error [1, 2]. The Royal London Hospital moved to a fully digital radiography imaging service in 2012 when a new building was opened. It was noted during 2012 that the number of internally reported radiation incidents attributed to either equipment or operator error increased, such as loss of images during transfer due to network problems and operators ignoring system warning messages. Reviewing these incidents highlighted the need for further operator training. More recently staff reporting or reviewing radiographs noted high exposure factors on a number of chest radiographs. A review of the doses and protocols was undertaken.

Journal of Physics: Conference Series **637** (2015) 012039 doi:10.1088/1742-6596/637/1/012039

2 Materials and Methods

2.1 Imaging equipment and radiography protocols

The general X-ray equipment in use at the Royal London Hospital is listed in Table 1. There are 20 Carestream DRX-1 wireless detectors, which can be moved between the different Carestream systems. The eight Carestream rooms are located in Accident and Emergency (A&E) X-ray, Inpatient X-ray and Outpatient X-ray. The two Xograph systems are mounted on a ceiling track and serve the Resuscitation Room in A&E. The mobile systems are allocated to different floors in the hospital.

Table 1. General X-ray equipment at the Royal London Hospital

System	Number	Detector	Converter	Pixel pitch (μm)
Carestream Evolution Dual Bucky	8	Carestream DRX-1	Gd_2O_2S	139
Carestream DRX Evolution mobile	5	Carestream DRX-1	Gd_2O_2S	139
Xograph Arco	2	Canon CXDI-70C	CsI	125
Shimadzu Dart mobile	3	Canon CXDI-50C	CsI	160

Typical modulation transfer function (MTF) and detective quantum efficiency (DQE) measurements for the three detector types measured during routine quality assurance (QA) testing are shown in Figures 1 and 2. The images required for the analysis were obtained at 70 kV, filtered with 1 mm Cu, with a detector entrance air kerma of approximately 10 μGy for the MTF measurement, and 2.5 μGy for the noise power spectrum measurement.

All the systems have programmed exposure factors and exposure charts for manual exposures.

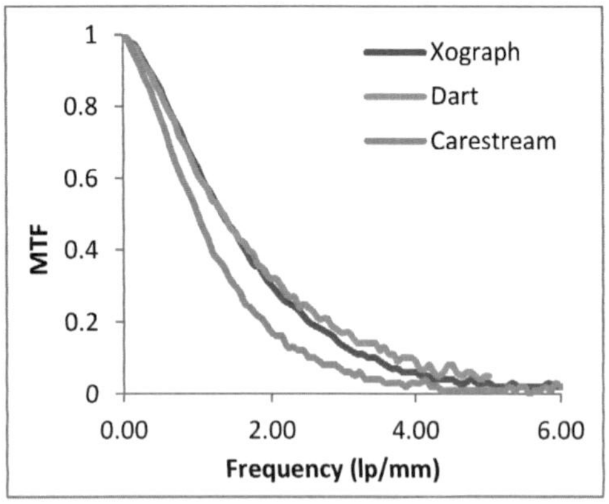

Figure 1. Typical MTF measurements for the 3 system types.

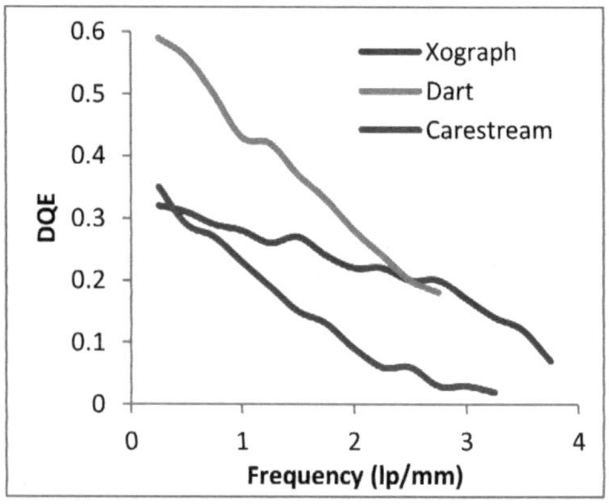

Figure 2. Typical DQE measurements for the 3 system types.

2.2 Dose Recording and Audit

Sectra PACS and RadCentre Radiology Information System (RIS) was used at the Royal London Hospital in 2014. Information on the X-ray examinations is entered into the RIS by the radiographer following the exposures and therefore is prone to data entry error. Examination type and room / mobile unit used are selected from drop down menus. Dose information such as Dose Area Product (DAP), kV and mAs is entered manually. The data extracted from the RIS system included the examination date, examination type, patient hospital ID, patient date of birth, X-ray room or system, kV, mAs, and DAP. Paediatric data was excluded from the data set. No information on patient weight or height was provided.

There were 37,352 adult chest X-ray examinations recorded on the RIS system at The Royal London Hospital in 2014. Of these 30,813 (82.5 %) had a DAP recorded. The room DAP calibration factor was then applied and all DAPs converted to $Gycm^2$. The first thing to consider was the accuracy of the data by applying cleansing to the RIS-based data, which was carried out by inspecting the DICOM headers of outliers. Multiple manual data entry errors were found including numbers of examinations entered under one examination code, data entered into incorrect fields, and incorrect DAP unit entered.

Due to the large number of manual data entry errors two approaches were used to review the data. Data was extracted from the DICOM headers via visual inspection of the headers for high dose examinations (DAP > 0.2 Gycm2, 650 examinations) and from a sample of 100 consecutive examinations in June 2014. For the second approach, patients who had multiple X-ray examinations during 2014 were reviewed.

3 Results and Discussion

3.1 Data sampled from June 2014

The mean DAP values for each X-ray system, including those in rooms and mobile units, were compared against the Local DRLs (LDRLs) and National DRLs (NDRLs) as shown in Figure 3. All rooms were below both the LDRLs and NDRLs, except for room RLH AEX5 (Resus). This room, which has an Xograph system, had only 4 examinations during the sampling period. All other rooms shown in the figure below are Carestream systems.

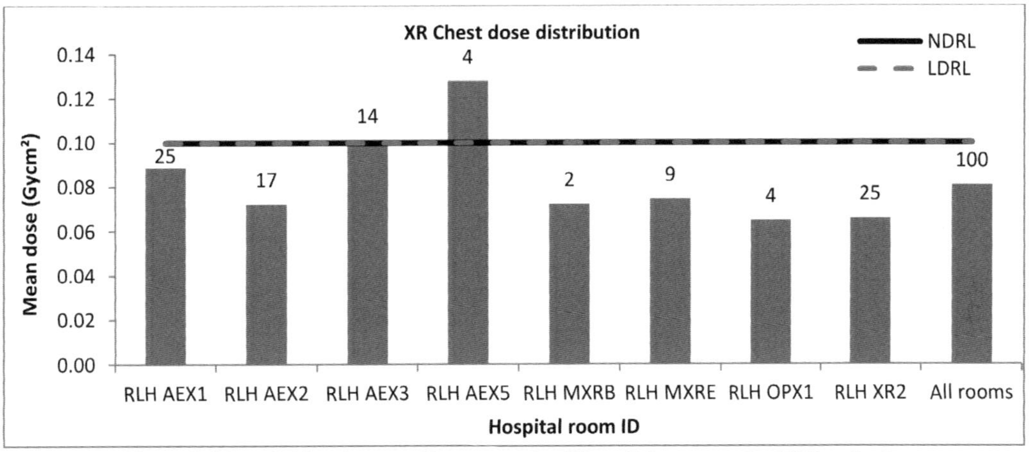

Figure 3. Mean DAP values per X-ray room/mobile unit for a sample of 100 patients.

3.2 High DAP data

Figures 4 and 5 show the DAP values > 0.2 $Gycm^2$ plotted against delivered mAs, and the distribution of delivered mAs respectively. Further patient records such as CT scans were inspected for the four outlying data point, and these were found to be bariatric patients. In Figure 4 the distribution of the delivered mAs is illustrated with one third of the examinations were exposed at 8 mAs, which is the backup setting for medium adults on the AEC.

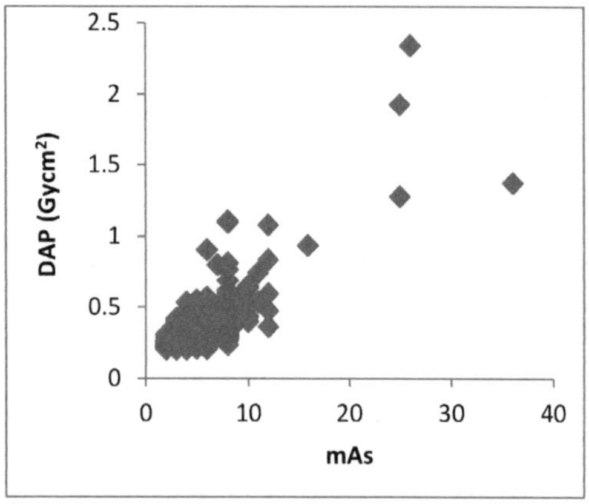

Figure 4. DAP values > 0.2 Gycm2 vs. delivered mAs.

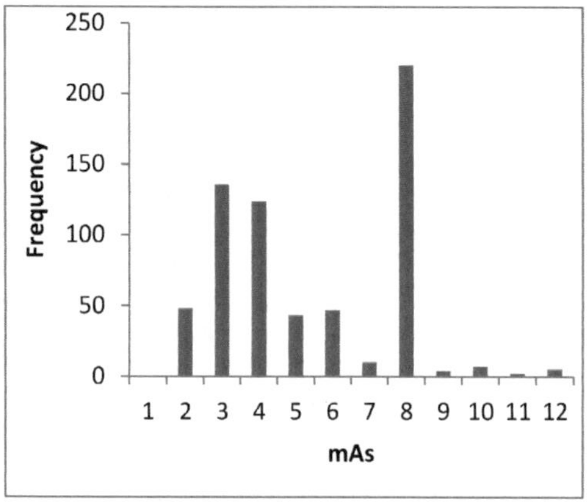

Figure 5. Distribution of the delivered mAs.

3.3 Example patient case study

A 68 year old female patient had 18 chest X-ray examinations in a two month period, of which 17 were examinations performed in a ward using a mobile X-ray system and therefore exposure parameters would have been selected manually. The selected kV, mAs and field sizes resulted in the DAP varying by an order of magnitude across these examinations. Table 2 summarises the exposure factors and DAP values.

Table 2. Exposure factors and DAP values for 18 examinations of one patient.

	Mean value	**Minimum**	**Maximum**
kV	86.8	85	90
mAs	2.6	1.6	4
DAP (Gycm^2)	0.10	0.03	0.22

4. Conclusions

The data presented demonstrate that there is a wide variation in the radiographic technique. The wide dynamic range of the detectors allows practices that could resulted in overexposure based radiography. These practices are not always identified and corrected.

The large number of examinations apparently terminated by the back-up timer were found not to be due to patient size or poor exposure factor selection, but due to the system defaulting to this setting under certain conditions, and this not being recognised prior to taking the exposure.

Radiographer training and audit is being undertaken to address this. The Trust is also in the process of implementing the Sectra DoseTrack management system on selected modalities, which will enable more effective management of these issues.

References

[1] Charnock, P., Moores, B. M. and Wilde, R. 2013. Establishing local and regional DRLs by means of electronic radiographical X-ray examination records. *Radiat. Prot. Dosim.* 157(1), pp 62-72.

[2] Wilde, R., Charnock, P., Moores, B. M. and McDonald, S. 2011. Qualifying the use of RIS data for patient dose by comparison with DICOM header data. *Radiat. Prot. Dosim.* 147(1-2), pp 329-332.

Tract-based Spatial Statistics and fMRI Analysis in Patients with Small Cell Lung Cancer before Prophylactic Cranial Irradiation

S Benezi[1], K Bromis[1], E Karavasilis[2], I S Karanasiou[1], M Koutsoupidou[1], G Matsopoulos[1], E Ventouras[3], N Uzunoglu[1], V Kouloulias[4], M Papathanasiou[4], A Foteineas[4], E Efstathopoulos[4], N Kelekis[4] and D Kelekis[2]

[1] National Technical University of Athens School of Electrical & Computer Engineering Microwave & Fiber Optics Laboratory.
[2] Research Centre of Radiology and Imaging, "Evgenidion" General Hospital.
[3] Department of Medical Instrumentation Technology, Technological Educational Institution of Athens.
[4] National and Kapodistrian University of Athens, Division of Radiology-Radiotherapy II

E-mail: sophiampenezi@gmail.com

Abstract. Prophylactic cranial irradiation (PCI) is known to increase life expectancy to a significant degree in Small Cell Lung Cancer (SCLC) patients. The overall scope of this research is to investigate changes in structural and functional connectivity between SCLC patients and controls before and after PCI treatment. In the current study specifically we use diffusion tensor imaging (DTI) and functional Magnetic Resonance (fMRI) to identify potential alterations in white matter structure and brain function respectively, in SCLC patients before PCI compared to healthy participants. The results in DTI analysis have showed lower fractional anisotropy (FA) and higher eigenvalues in white matter regions in the patient group. Similarly, in fMRI analysis a lower level of activation in the primary somatosensory cortex was reported. The results presented herein are subject to further investigation with larger patient and control groups.

1. Introduction

Prophylactic cranial irradiation (PCI) is a standard method to prevent metastatic tumors in the brain for small cell lung cancer patients. Despite the fact that PCI prolongs life expectancy, negative effects (such as chemotherapy) in various cognitive functions have been reported [1].

In order to discriminate those effects, Diffusion Tensor Imaging (DTI) has been applied. DTI can detect white matter (WM) lesions [2] by measuring the diffusion of water molecules in the brain. The most common parameters of this technique are mean diffusivity (MD) and fractional anisotropy (FA), which define the magnitude of diffusion and the preferential directionality of water diffusion along the white matter tracts, respectively [3].

Journal of Physics: Conference Series **637** (2015) 012040 doi:10.1088/1742-6596/637/1/012040

Additionally, functional Magnetic Resonance Imaging (fMRI) technique has been implemented in order to determine the null hypothesis of neurocognitive impairments. fMRI measures brain activity by detecting changes in the local oxygenation of blood. In the current study, we employed two fMRI techniques; task-related and resting state fMRI. During rest, several brain regions are functionally connected thus forming resting-state networks. These networks are of great importance for the characterization of the complex patterns of neural and behavioural consequences of cancer.

In the present study, having a larger sample size and more statistical power than in previous attempts [4], [5], [6], [7] we sought to examine for potential differentiations by comparing SCLC patients with normal controls using DTI as well as fMRI techniques.

2. Methods

Thirteen healthy participants and seventeen SCLC patients before PCI treatment participated in the study. All participants had complete response to initial treatment (chemotherapy with or without chest radiotherapy), without any brain metastases shown, met the standard MRI safety criteria and had no history of diagnosed neurological disorder, major psychiatric disorder or treatment with psychotropic medication, including substance misuse. Whole brain MRI, DTI and fMRI data were collected on a Philips 3.0T scanner (Achieva; Philips, Best, The Netherlands) at the Radiology Research Unity, Medical Imaging Department, Evgenidion Hospital, National and Kapodistrian University, Athens, Greece using an 8-channel SENSE head coil used for radiofrequency reception of the nuclear magnetic resonance signals. Foam pads and headphones were used to reduce head motion and scanner noise. Anatomical imaging was performed with T1-weighted 3D sagittal acquisition (1.0-mm-thick slices, 0 mm slice gap, TE = 4.6 msec/TR = 15 msec, FOV = 256, and 1.0 x 1.0 x 1.0 mm^3 reconstructed voxel size, and T2 Fluid Attenuated Inversion Recovery (FLAIR) acquisition (1.0-mm-thick slices, 0 mm slice gap, TE = 4.6 msec/TR = 15 msec, FOV = 256, and a 1.0 x 1.0 x 1.0 mm^3 reconstructed voxel size.

All data were preprocessed and analyzed using FSL (FMRIB's Software Library, version 5.0;www.fMRIb.ox.ac.uk/fsl) [8]. Images from the whole dataset were initially brain-extracted using BET (Brain Extraction Tool) [8].

DTI data set was corrected for stretches and shears induced by eddy currents in the gradient coils and simple head motions by using affine transformation. A diffusion tensor model was fitted on the data to determine the level of anisotropy for each voxel independently. Fractional anisotropy, λ1, λ2 and λ3 maps were generated using FDT-FMRIB's Diffusion Toolbox [9]. The FA was calculated and plotted in a single FA map for each subject. Voxel-wise statistical analysis of the FA data was performed using TBSS implemented in FSL [10]. TBSS analysis was also applied to the λ1, λ2 and λ3 maps, as well as in MD.

fMRI data were realigned for motion correction (MCFLIRT) [11]. A high-pass filtering was implemented (60 s cutoff) and images were spatially smoothed using a Gaussian kernel of 6 mm full width at half maximum. Artefacts were removed using MELODIC (Independent Component Analysis-ICA tool in FSL). The median functional image was aligned to the high-resolution T1-weighted image using a rigid body transformation [11] and then registered to the T1 MNI152 template using affine as well as non-linear transformations with a warp resolution of 10mm. First level analysis was conducted using FILM (FMRIB's Improved Linear Model) [12]. The contrasts of interest were left-tapping vs. baseline and right-tapping vs. baseline. Group comparisons were conducted using mixed effects (ME) analysis. Finally, two-sample unpaired t-tests were accomplished in order to estimate each group's cross-subject variability separately ($p<0.05$).

Resting state fMRI data were decomposed into 42 independent components and analyzed using multi-session temporal concatenation group ICA (MELODIC FSL tool). Model order was estimated using the Laplace approximation to the Bayesian evidence for a probabilistic principal component model. For each subject, the group-average spatial maps were regressed onto the subject's 4D space-time dataset (spatial regression) using a regression technique called dual regression. A set of subject-specific timeseries were created, one per group-level spatial map. Those timeseries were regressed into the same 4D dataset, resulting in a subject-specific spatial correlation maps. The component representing the default mode network (DMN) was identified visually [13]. Group comparisons were conducted in this network by performing nonparametric permutation tests (5000 permutations) using FSL's randomize permutation-testing tool. Additionally, to control the probability of false activations, a family-wise error correction (FWE) was performed using a threshold of $p<0.05$.

3. Results

Results from DTI analysis report that brain regions within the centrum semiovale WM, the cerebellum as well as the fornix may represent lower FA in SCLC patients compared to controls (voxelwise thresholding uncorrected for multiple comparisons, $t>3$, $p<0.05$). Furthermore, TBSS analysis showed that regions with FA reductions were accompanied by increased λ1, λ2, λ3 and MD (figure 1).

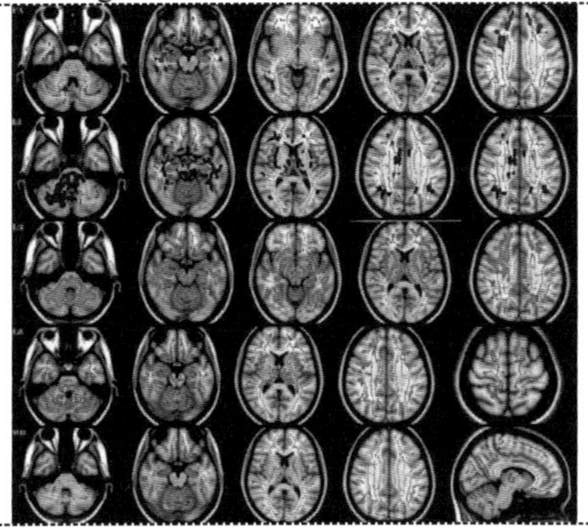

Figure 1. Significant differences, where patient FA is lower than control FA, are visualized with red coloring and the increase of λ1, λ2, λ3 and MD are visualized with blue, purple, yellow and cool respectively. Mean FA skeleton is presented in green color. The first row shows differences in FA, the second in λ1, the third in λ2, the fourth in λ3 and the last row in MD.

Results from task-related fMRI analysis showed common activations in both patients and healthy controls (including the premotor cortex, the primary motor cortex, the primary somatosensory cortex and the corticospinal tract) during the left and right tapping task. Additionally, when comparing the two groups, the healthy participants showed a significantly higher level of activation in the left as well as right primary somatosensory cortex (figure 2). Resting state fMRI analysis results showed that for both groups the DMN was significantly activated (figure 3).

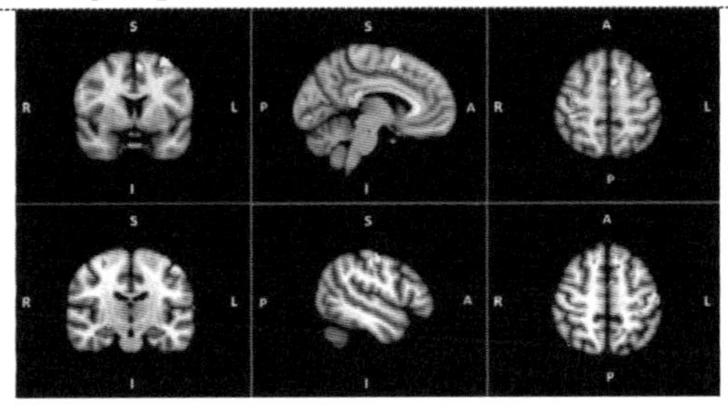

Figure 2. Comparison between the control and patient groups for the left tapping (top images) and the right tapping (bottom images).

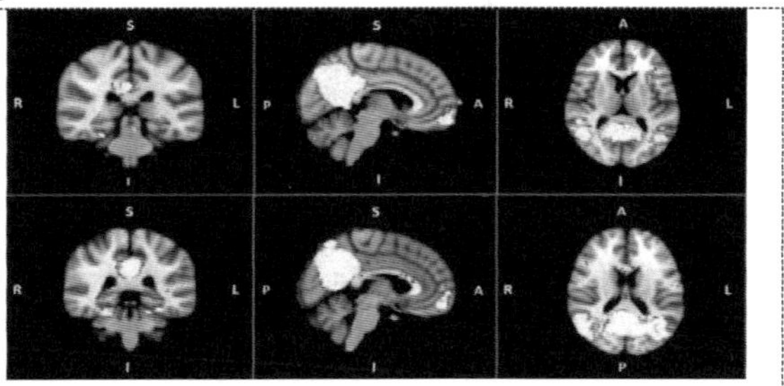

Figure 3. Default Mode Network in healthy group (top images) and patient group (bottom images).

4. Discussion and Conclusion

In this study, we found some WM tract damages in patients with SCLC before PCI treatment. The results show decreased FA in the centrum semiovale WM, the cerebellum and the fornix. Furthermore, three eigenvectors of the diffusion tensor and the MD were analyzed. λ2 and λ3 were increased while λ1 and MD revealed smaller changes. In terms of fMRI analysis, when comparing the two groups, the healthy participants showed a significantly higher level of activation in the left and right primary somatosensory cortex. Moreover, during rs-fMRI, the DMN was identified in both groups.

As this research is ongoing, anatomical and functional connectivity in healthy subjects and patients with small cell lung cancer [4], [5], [6], [7] depending on the type of disease and the phase of treatment and time of clinical examination, will be studied. Study of cognitive functionality of patients with SCLC will be conducted with imaging and clinical examination. The effects of radiation (PCI) both at anatomical as well as functional levels in the brain will be consequently assessed. Efforts are being made to increase the number of participants at different stages of disease and treatment which is a difficult task taking into consideration the significant percentage of mortality of this specific type of cancer patients.

5. References

[1] Kanard A, Frytak S, Jatoi A. Cognitive dysfunction in patients with small-cell lung cancer: incidence, causes, and suggestions on management. J Support Oncol. 2004 Apr;2(2):127–32; discussion 133–5, 138–40.

[2] Chapman CH, et al. Regional Variation in Brain White Matter Diffusion Index Changes following Chemoradiotherapy: A Prospective Study Using Tract-Based Spatial Statistics. PLoS ONE. 2013;8(3):e57768.

[3] Aliotta R, et al. Tract-based spatial statistics analysis of diffusion-tensor imaging data in pediatric- and adult-onset multiple sclerosis. Hum Brain Mapp. 2014 Jan;35(1):53–60.

[4] Bromis K, et al. Analysis of resting state and task-related fMRI data in small cell lung cancer patients before undertaking PCI. 2014 EAI 4th International Conference on Wireless Mobile Communication and Healthcare (Mobihealth). 2014. p. 91–4.

[5] Bromis K, , et al. Resting state and task related fMRI in small cell lung cancer patients. 2013 IEEE 13th International Conference on Bioinformatics and Bioengineering (BIBE). 2013. p.1–4.

[6] Benezi S, et al. Tract-Based Spatial Statistics Analysis of Diffusion-Tensor Imaging Data in Patients with Small Cell Lung Cancer. ICST; 2014 http://eudl.eu/doi/10.4108/icst.mobihealth.2014.257513

[7] Benezi SE, Karavasilis et al. Tract – based spatial statistics in patients with small cell lung cancer. Physica Medica. 2014;30:e87.

[8] Smith SM, et al. Advances in functional and structural MR image analysis and implementation as FSL. Neuroimage. 2004;23 Suppl 1:S208–19.

[9] Behrens TEJ, et al. Characterization and propagation of uncertainty in diffusion-weighted MR imaging. Magn Reson Med. 2003 Nov;50(5):1077–88.

[10] Smith SM, , et al. Tract-based spatial statistics: voxelwise analysis of multi-subject diffusion data. Neuroimage. 2006 Jul 15;31(4):1487–505.

[11] Jenkinson M, Bannister P, Brady M, Smith S. Improved optimization for the robust and accurate linear registration and motion correction of brain images. Neuroimage. 2002 Oct;17(2):825–41.

[12] Damoiseaux JS, et al. Consistent resting-state networks across healthy subjects. PNAS. 2006 Sep 12;103(37):13848–53.

[13] Beckmann CF, DeLuca M, Devlin JT, Smith SM. Investigations into resting-state connectivity using independent component analysis. Philos Trans R Soc Lond, B, Biol Sci. 2005 May 29;360(1457):1001–13.

6. Acknowledgment

This research has been co-financed by the European Union (European Social Fund – ESF) and Greek national funds through the Operational Program "Education and Lifelong Learning" of the National Strategic Reference Framework (NSRF)-Research Funding Program: THALIS – NTUA, "Study and Analysis of Medical Data using structural and functional Magnetic Resonance Imaging procedures (MRI/DTI/fMRI): Assessment of changes induced by Brain Radiotherapy" (MIS 380151).

Image guided Brachytherapy: *The paradigm of Gynecologic and Partial Breast HDR Brachytherapy*

S Diamantopoulos[1], I Kantemiris[1], A Konidari[1], P Zaverdinos[1]
[1]Medical Physics Department, Metropolitan Hospital, N. Faliro, Greece

E-mail: pzaverdinos@metropolitan-hospital.gr

Abstract. High dose rate (HDR) brachytherapy uses high strength radioactive sources and temporary interstitial implants to conform the dose to target and minimize the treatment time. The advances of imaging technology enable accurate reconstruction of the implant and exact delineation of high-risk CTV and the surrounding critical structures. Furthermore, with sophisticated treatment planning systems, applicator devices and stepping source afterloaders, brachytherapy evolved to a more precise, safe and individualized treatment. At the Radiation Oncology Department of Metropolitan Hospital Athens, MRI guided HDR gynecologic (GYN) brachytherapy and accelerated partial breast irradiation (APBI) with brachytherapy are performed routinely. Contouring and treatment planning are based on the recommendations of the GEC – ESTRO Working group. The task of this presentation is to reveal the advantages of 3D image guided brachytherapy over 2D brachytherapy. Thus, two patients treated at our department (one GYN and one APBI) will be presented. The advantage of having adequate dose coverage of the high risk CTV and simultaneous low doses to the OARs when using 3D image-based brachytherapy will be presented. The treatment techniques, equipment issues, as well as implantation, imaging and treatment planning procedures will be described. Quality assurance checks will be treated separately.

Keywords: HDR Brachytherapy, GYN, APBI

1. Introduction

The introduction of brachytherapy dates back to the beginning of the last century. During brachytherapy energy is derived from a radioactive source which is located in the proximity of the tumour. Therefore, radiation is deposited directly into the tumour and radiation induced side effects are minimized because of the high conformity of the procedure [1].
Modern brachytherapy exploits technological advantages such as high strength sealed radioactive sources, remotely controlled afterloading machines, sophisticated treatment planning systems and temporary implants. High doses are delivered within short treatment times. The latter is called "High Dose Rate" (HDR) brachytherapy.
Based on the technique of applicator placement HDR brachytherapy can be divided into two main subgroups: a) intracavitary and b) interstitial. During intracavitary brachytherapy specific applicators are inserted into cavities (e.g. gynecological brachytherapy). Interstitial brachytherapy is used if the tumour is not accessible through cavities. Hollow needles are then implanted into the tissue, using adequate distributions in order to cover the malignant site. Although there is a clear distinction between these two types, it is common practice to combine both techniques if this is required in order to maximize target coverage [1].

Image guidance plays a substantial role for modern brachytherapy which represents a highly conformal treatment. The doses delivered to the target are significantly higher compared to those delivered with external beam radiation therapy (EBRT). Therefore, it is of major importance to precisely reconstruct the implant, as well as to delineate the high risk clinical target volume (HR-CTV) and the surrounding healthy organs with high accuracy.
Imaging innovations which were developed during the past decades were integrated into the procedure of brachytherapy (e.g. Ultrasound (US), Computed Tomography (CT), Magnetic Resonance Imaging (MRI)). MRI with its superior soft tissue depiction quality represents the gold standard for GYN brachytherapy. US is utilized mainly for prostate brachytherapy and CT for many different sides [2].
At the Radiation Oncology Department of the Metropolitan Hospital Athens, MR Image Guided Adaptive BrachyTherapy (IGABT) for GYN tumours and Accelerated Partial Breast Irradiation (APBI) with interstitial brachytherapy are performed on a routine basis. Both treatments are very demanding with regard to equipment, applicator implantation, imaging, treatment planning and quality assurance checks.
The present work provides a brief description for both techniques, focusing mainly on technical aspects.

2. Materials & Methods

At the Radiation Oncology Department of the Metropolitan Hospital MR IGABT for GYN tumours is performed on a daily basis. Applicator sets, catheters and needles which are used are provided by Elekta company (Elekta A.B. Stockholm, Sweden). For imaging purposes, a CT – SIM (Somatom CT scanner (Siemens Medical Solutions USA, Inc, Malvern, USA)) and a MRI scanner (Magnetom Avanto (Siemens Medical Solutions USA, Inc., Malvern, USA)) are utilized. Treatment planning is perfomed with Oncentra Brachy system, v. 4.5 (Elekta A.B. Stockholm, Sweden). Treatment is performed with a microSelectron v.2. afterloader (Elekta A.B. Stockholm, Sweden). On a scheduled basis (daily, monthly and yearly) dosimetric checks of the brachytherapy equipment are performed using Well Type Chamber, EBT films, Electa's check ruler, Survey Meter and chronometer.

2.1. Gynecological Brachytherapy – Cervical Cancer

Cervical cancer brachytherapy can be performed either as a standalone treatment instead of surgery for small (IA) tumours, or in combination with EBRT and chemotherapy for more advanced tumours [1].
A combined intracavitary/interstitial tandem-ring applicator (Vienna applicator) [3], which is MRI compatible, is frequently used for this kind of brachytherapy. The ability to adapt to each individual anatomical and pathological situation is provided by different ring diameters (26, 30, 34 mm) and different tandem-intrauterine lengths (2, 4, 6 cm). In addition, each ring has on its periphery a set of holes, where specific Titanium needles can be added, for adequate dose coverage to the target.
The insertion of the implant is performed with the patient in lithotomy position under spinal anaesthesia. A Foley catheter is inserted into the bladder and is filled with a radiation-opaque fluid. A transrectal US probe assists for the correct insertion of the entire implant.
Following implantation, CT and MR images are acquired. T2 weighted images in three planes are acquired for target and organ delineation purposes. T1 weighted images are obtained for applicator reconstruction. Images are imported into the treatment planning system in DICOM format. The delineation of the OARs (bladder, rectum, sigmoid) and the definition of the HR CTV is completed by the radiation oncologist. Subsequently, the medical physicist reproduces "in silico" the exact implant, either by using vendor provided libraries of applicator models or by using manual reconstruction (Figure 1).
Although in an APBI implant, Paris system is followed, sometimes the needles are curved and are not equidistant as it is shown in Fig. 2. In 3D brachytherapy these "imperfections" of the implant can be corrected – in contrary to the 2D brachytherapy – by dwell time optimization.

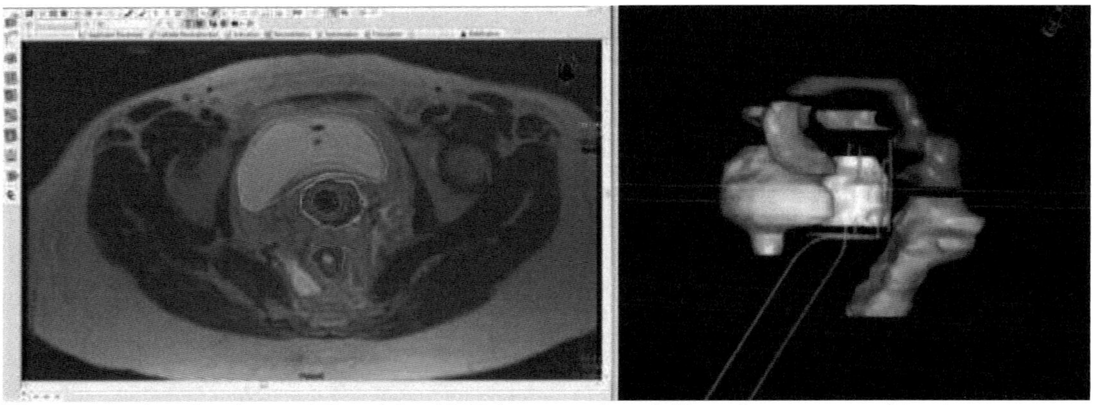

Figure1. Isodose levels on a T2 weighted MR image and model based reconstructed implant (tandem-ring and peripheral catheters) for cervical brachytherapy.

The goal of the planning procedure is to shape the dose distribution of the implant to the target volume, using source position and dwell time optimization. The prescription dose is 7 Gy per fraction, four fractions in total. Dosimetric goals for this treatment are D90 (The dose that encompass at least the 90% of the high-risk CTV [1]]) ≥ 7Gy, D2cc of bladder (the dose that at least receive the 2 cm^3 of the bladder [1]) < 5.1 Gy, D2cc of rectum (the dose that at least receive the 2 cm^3 of the rectum [1]) ≤ 4.8 Gy and D2cc of sigmoid the dose that at least receive the 2 cm^3 of the sigmoid [1]) ≤ 4.8 Gy. Fig. 1 demonstrates conformity of MR IGABT: Invasion of the tumour into the left parametria was covered with the insertium of titanium needles which were appropriately optimized.

2.2. Accelerated Partial Breast Irradiation using Interstitial Brachytherapy
Accelerated partial breast brachytherapy is offered to patients as an alternate treatment option to EBRT [4, 5]. Patient selection criteria are age (> 50years), histology (grade 1-3) and tumour size (< 3cm). Needle implantation procedure is performed either in an intraoperative or postoperative manner. The needles are placed into the tumour bed aiming at parallel and equidistant distributions according to the Paris system [1] (Figure 2).

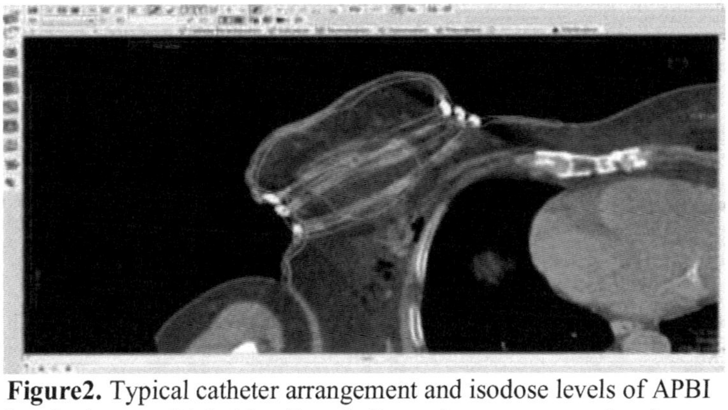

Figure2. Typical catheter arrangement and isodose levels of APBI brachytherapy. Light blue lines indicate the reconstructed catheters.

After surgery a CT scan is performed. DICOM images are imported into the treatment planning system. The radiation oncologist delineates the target and the organs at risks (skin, underlying ribs, lung and remaining breast).

Treatment planning is based on the rules of Paris system for interstitial brachytherapy. Planning target volume (PTV) should be covered by the 85% isodose level of the mean central dose. Prescription dose is 32Gy in 8 fractions of 4Gy twice daily. Dosimetric goals are presented in Table1.

Table 1. Minimum quality requirements for plan acceptance in an APBI treatment.

Dosimetric Indexes	**Minimum Values**
D90	$\geq$ 90%
V100	> 90%
DNR[a]	$\leq$ 0.30
D_{skin}	< Prescribed dose
D_{lung}	< 70% Prescribed dose

[a] Dose nonconformity ratio-DNR is the ratio of the volume of 150% isodose to the 100% isodose

3. QA protocol for HDR brachytherapy

Prior to each brachytherapy treatment several safety Quality Assurance (QA) checks are carried out. The list of QA checks includes daily routine tests for verification of the impeccable performance of the equipment, as well as patient specific measurements for precise implant reconstruction and reproducibility between fractions.
On a daily basis, the current activity of the Iridium 192 source is calculated and compared with treatment planning and afterloader values. Source position checks and time checks are also performed, along with numerous equipment safety checks: communication, catheter attachment lock, door interlock, warning lights, room monitor, emergency stop buttons, catheter obstruction, treatment timer and contamination tests [6].
Some of the patient specific checks are: measurement of the free length of the implanted catheters, comparison of the active dwell times derived from plan to the actual that are going to be delivered, test for possible source obstruction in the catheters.

4. Conclusions

Modern image based adaptive HDR brachytherapy is a highly demanding procedure. It represents a high end technique which utilizes sophisticated equipment and imaging, while expertise plays a pivotal role. Finally, it is of major importance, that the medical physics team of the department enables safe and accurate dose delivery.

5. References

[1] V. Strnad, R. Potter and G. Kovacs, Practical Handbook of Brachytherapy (2014).
[2] J.C. Dimopoulos, G. Schirl, A. Baldinger, T.H. Helbich and R. Potter, MRI assessment of cervical cancer for adaptive radiotherapy, *Strahlenther Onkol* **185** (2009), pp. 282- 287.
[3] C. Kirisits, S. Lang, J. Dimopoulos, D. Berger, D. Georg and R. Potter, The Vienna applicator for combined intracavitary and interstitial brachytherapy of cervical cancer: design, application, treatment planning, and dosimetric results, *Int J Radiat Oncol Biol Phys* **65** (2006), pp. 624-630.
[4] P. Karlsson, Accelerated partial breast cancer irradiation (APBI)--the future breast cancer radiotherapy?, *Acta Oncol* **48** (2009), pp. 485-486.
[5] F. Wenz, W. Budach, J. Dunst, P. Feyer, W. Haase, W. Harms, *et al.*, Accelerated partial breast irradiation (APBI)--ready for prime time?, *Strahlenther Onkol* **185** (2009), pp. 653-655.
[6] J. Venselaar and P.-C. J, A practical quide to quality control of brachytherapy equipment (2004).

Shielding of Sensitive Electronic Devices in Magnetic Nanoparticle Hyperthermia Using Arrays of Coils

S V Spirou, P Tsialios and G Loudos

Department of Biomedical Engineering, Technological Educational Institute of Athens, Athens, Greece

E-mail: spiridonspirou@yahoo.com

Abstract. In Magnetic Nanoparticle Hyperthermia (MNH) an externally applied electromagnetic field transfers energy to the magnetic nanoparticles in the body, which in turn convert this energy into heat, thus locally heating the tissue they are located in. This external electromagnetic field is sufficiently strong so as to cause interference and affect sensitive electronic equipment. Standard shielding of magnetic fields involves Faraday cages or coating with high-permeability shielding alloys; however, these techniques cannot be used with optically sensitive devices, such as those employed in Optical Coherence Tomography or radionuclide imaging. In this work we present a method to achieve magnetic shielding using an array of coils. The magnetic field generated by a single coil was calculated using the COMSOL physics simulation toolkit. Software was written in C/C++ to import the single-coil data, and then calculate the positions, number of turns and currents in the shielding coils in order to minimize the magnetic field strength at the desired location. Simulations and calculations have shown that just two shielding coils can reduce the magnetic field by 2-3 orders of magnitude.

1. Introduction

Hyperthermia (HT) in Oncology refers to the artificial increase of local temperature for the treatment of cancer, typically in the range of 41° C to 45° C for 1 hour or more [1, 2]. Although hyperthermia alone does not seem to have a curative effect, there is significant evidence that a strong synergistic effect is achieved when HT is combined with other treatment modalities. With Radiation Therapy (RT), in particular, HT acts as a powerful radiosensitizer, especially addressing the weaknesses of RT, such as hypoxic regions and cells in radioresistant phases [3-5]. In addition, the abnormal vasculature of the tumor is unable to respond to the external heat the way normal blood vessels do, i.e. increasing blood flow in order to cool the tissues involved, so that the effect of HT remains localized in the tumor [2, 3].

HT has been traditionally administered using radiofrequency, microwave, ultrasound and infrared radiation techniques. In recent years, magnetic nanoparticles have emerged as a possible alternative to these techniques. In Magnetic Nanoparticle Hyperthermia iron-oxide nanoparticles in the form of an aqueous solution are directly injected in the tumor and heated by an externally applied time-varying magnetic field [6, 7]. The strength of the magnetic field required to achieve the desired thermal effect, typically 10^4 kA/m, is sufficiently high so as to cause interference in nearby electronic devices. Standard shielding of magnetic fields involves Faraday cages or coating with high-permeability shielding alloys; however, these techniques cannot be used with optically sensitive devices, such as those employed in Optical Coherence Tomography or radionuclide imaging. In this work we present a method to achieve magnetic shielding using an array of coils or solenoids.

2. Methods

The basic setup is illustrated in figure 1. All the coils are coaxial. The HT coil is held fixed at $x = 0$ and the shielding coils are placed between the HT coil and the imaging device in order to reduce the magnetic field $\boldsymbol{B}$ at the desired location where the sensitive electronics lie.

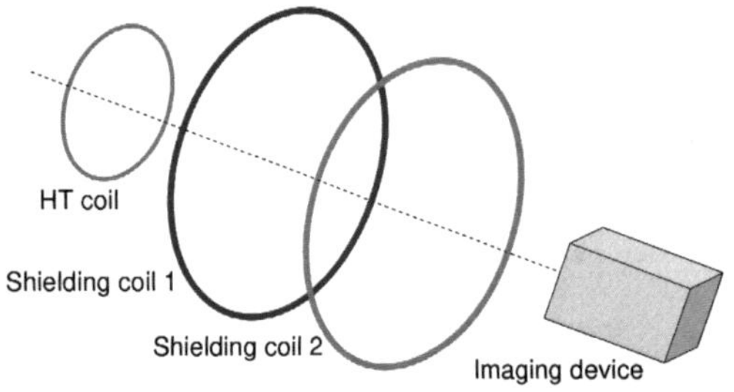

Figure 1. The experimental setup showing the HT and shielding coils, and the magnetically sensitive imaging device.

Two scenarios were investigated: in the *small*-size scenario the dimensions are similar to those found in a typical small-animal experiment, whereas in the *large*-size scenario the dimensions are similar to those found in a typical large-animal or human experiment. In both scenarios both the HT and shielding coils consist of 10 turns each and are 5 cm in length and thickness. The current through the HT coil is held fixed at 100 mA. The remaining parameters depend on the scenario and are listed in table 1.

Table 1. Coil parameters that vary between the scenarios investigated.

Scenario	HT radius (cm)	Shielding coil 1		Shielding coil 2	
		Radius (cm)	Location (cm)	Radius (cm)	Location (cm)
Small	10	20	10	20	20
Large	40	80	100	80	200

The axial field of a thin, finite coil in the SI system is given by [8,9,10]:

$$B = \frac{\mu_0 IN}{2L}\left[\frac{x_2}{\sqrt{x_2^2+r^2}} - \frac{x_1}{\sqrt{x_1^2+r^2}}\right] \quad (1)$$

where μ_0 is the permeability constant (1.26 x 10^{-6} H/m), I is the current in the coil, N is the total number of turns, L is the length of the coil, r is the radius of the coil, and x_1 and x_2 are the distances from the two ends of the coil.

For a solenoid of finite thickness (1) becomes [8,10,11]:

$$B = \frac{\mu_0 IN}{2(r_2-r_1)}\left[x_2 \ln\frac{\sqrt{r_2^2+x_2^2}+r_2}{\sqrt{r_1^2+x_2^2}+r_1} - x_1 \ln\frac{\sqrt{r_2^2+x_1^2}+r_2}{\sqrt{r_1^2+x_1^2}+r_1}\right] \quad (2)$$

where r_1 and r_2 are the inner and outer radii of the coil.

Alternatively, the field of each coil can be calculated using finite elements techniques, such as the COMSOL physics simulation toolkit (www.comsol.com). The advantage of this approach is that the calculation produces the ***B*** field in full 3-dimensional space, not just along the axis of the coils. The ***B*** field can then be exported for further use.

If inductive currents and fields and other interplay effects between the coils are ignored, then the resulting field is a superposition of the field generated by each coil separately. Software was written in C++ to calculate the total axial field, using the data calculated by COMSOL.

The shielding effect was quantified using the shielding factor, which is defined as:

$$\textit{Shielding Factor} = \log_{10}\left[\frac{\mathbf{B}\,\text{without shielding}}{\mathbf{B}\,\text{with shielding}}\right] \tag{3}$$

It should be noted that since ***B*** is proportional to the product of the number of turns times the current through the coil, for the purposes of this investigation it suffices to keep the number of turns constant and modify only the current. In an actual implementation, one might choose to change the number of turns in order to increase or decrease the absolute magnitude of the current.

3. Results

The relative magnitude of ***B***, normalized to the maximum value when only the HT coil is present, is shown in figures 2 and 3 for the small and large-size scenarios, respectively. In the large-size scenario the currents through the two shielding coils are −3.5 and −0.0253 mA (the minus sign indicates that the current flows in the opposite direction from the current in the HT coil). For $x >= 2.5$ m the field is uniform and the shielding factor is 2.6–2.8.

In the small-size scenario with one shielding coil, the current through the shielding coil is −14 mA. A uniform field is achieved for $x >= 46$ cm with a shielding factor of approximately 1.8–1.9. When two shielding coils are used, the current through the first one is −14.45 mA and the current through the second one is 0.11 mA. The resulting field is not uniform; however, the shielding effect is much higher and it begins closer to the HT coil. In the range $41 \leq x \leq 57$ cm the shielding factor is no less than 2.6. At $x = 42.5$ and 52.1 cm the shielding factor increases to 3.5 and 4.4, respectively.

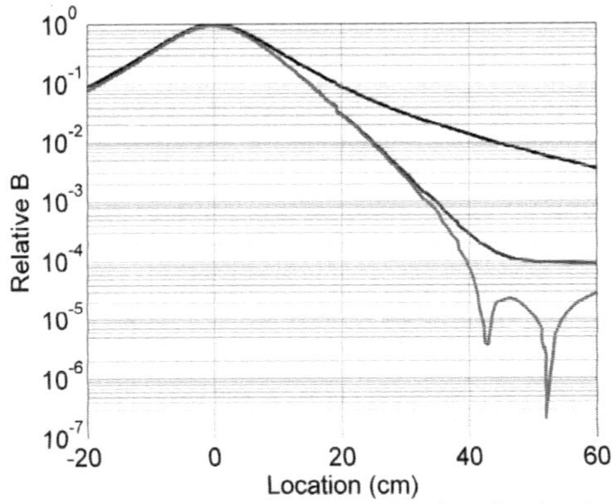

Figure 2. The relative magnitude of B in the small-size scenario: The HT coil alone (black), with one shielding coil (blue), with two shielding coils (red)

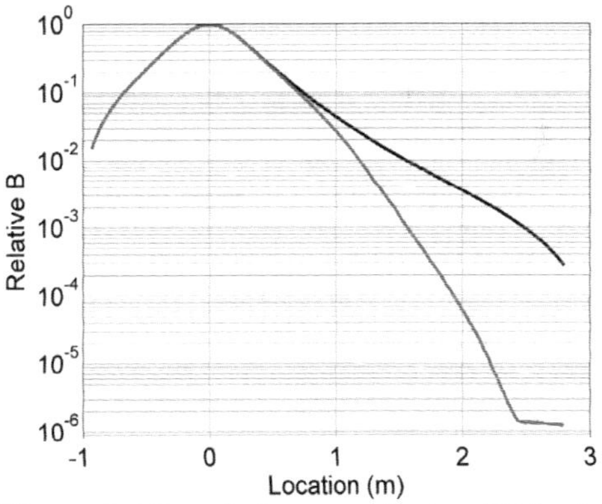

Figure 3. The relative magnitude of B in the large-size scenario: The HT coil alone (black), with two shielding coils (red)

4. Discussion and Conclusion

We have presented a method to reduce the magnetic field at a given distance from the hyperthermia coil, without affecting the optical properties of the imaging device. In the large-size scenario, a uniform field is achieved using two shielding coils, both of whose currents flow in the opposite direction from the current in the HT coil. In the small-size scenario, a uniform field is achieved using a single shielding coil. The addition of a second coil may significantly increase the shielding factor, by more than an order of magnitude at specific points, at the expense of creating a non-uniform resulting field. Furthermore, it causes the shielding effect to begin closer to the HT coil. Both these characteristics are important since: (a) depending on the construction of the imaging device, the sensitive electronics may be only a few mm in size, and (b) in general, the imaging device, and thus the sensitive electronics, should be as close as possible to the organ or object to be imaged.

Acknowledgement

This research has been co-funded by the European Union (European Social Fund) and Greek national resources under the framework of the “ARISTEIA II” project HYPERGNOSTIC code 4309 of the “Education & Lifelong Learning” Operational Programme.

References

[1] Sneed P K, Stauffer P R, Li G C, Sun X and Myerson R J 2010 Hyperthermia *Leibel’s textbook of radiation oncology* ed R T Hoppe, T L Phillips and M Roach (Philadelphia, PA: Saunders)
[2] Chichel A, Skowronek J, Kubaszewska M and Kanikowski M 2007 Hyperthermia – description of a method and a review of clinical applications *Rep Pract Oncol Radiother* **12** 267-75
[3] Horsman M R and Overgaard J 2007 Hyperthermia: a potent enhancer of radiotherapy *Clin Oncol* **19**: 418-26
[4] Horsman M R and Overgaard J. 2002 Overcoming tumour radioresistance resulting from hypoxia. *Basic clinical radiobiology for radiation oncologists*, ed G G Steel (London: Edward Arnold) 169-181.
[5] Dewey W C, Hopwood L E, Sapareto S A and Gerweck L E 1977 Cellular responses to combinations of hyperthermia and radiation *Radiology* **123** 463-74
[6] Dutz S and Hergt R 2014 Magnetic particle hyperthermia – a promising tumour therapy? *Nanotechnology* 25 452001
[7] Jordan A 2005 Thermotherapy using magnetic nanoparticles: Principles and clinical application of nanotherapy *EuroNanoForum Proceedings*, pp. 76–80
[8] Jackson J D 1999 *Classical Electrodynamics* (John Wiley and Sons)
[9] Callaghan E E and Maslen S H 1960 The magnetic field of a finite solenoid *NASA Technical Note D-465* (Washington, DC: NASA)
[10] http://www.netdenizen.com/emagnet/index.htm
[11] Brown G V, Flax L, Itean E C and Lawrence J C 1963 Axial and radial magnetic fields of thick, finite-length solenoids *NASA Technical Report R-170* (Washington, DC: NASA)

Modelling and Simulation of the Knee Joint with a Depth Sensor Camera for Prosthetics and Movement Rehabilitation

S. Risto and M. Kallergi

Department of Biomedical Engineering, Technological Educational Institute of Athens, Athens 12243, Greece

E-mail: stamtherist@hotmail.com

Abstract. The purpose of this project was to model and simulate the knee joint. A computer model of the knee joint was first created, which was controlled by Microsoft's Kinect for Windows. Kinect created a depth map of the knee and lower leg motion independent of lighting conditions through an infrared sensor. A combination of open source software such as Blender, Python, Kinect SDK and NI_Mate were implemented for the creation and control of the simulated knee based on movements of a live physical model. A physical size model of the knee and lower leg was also created, the movement of which was controlled remotely by the computer model and Kinect. The real time communication of the model and the robotic knee was achieved through programming in Python and Arduino language. The result of this study showed that Kinect in the modelling of human kinematics and can play a significant role in the development of prosthetics and other assistive technologies.

1. Introduction

There are millions of people worldwide living with a lower limb loss, about 2 million of them in the United States alone.[1] The main causes of a limb loss are disease, such as diabetes and peripheral arterial disease, trauma, and congenital deformities.[2] A prosthesis is a common solution to upper and lower limb amputation and it may be either below or above the joint, i.e., the elbow or the knee. The creation of a model for the knee joint and simulation of the knee movement is an essential step in prosthesis design and calibration, movement rehabilitation, or diagnosis of human knee disabilities. Such model requires a multidisciplinary approach from physiology to computer science and mechanical and electrical engineering.

Microsoft's Kinect sensor has already been used in combination with computer vision algorithms to analyse motion patterns of healthy humans and amputees in an effort to improve prostheses design and other treatments.[3] Kinect offers low cost, computationally inexpensive motion and voice recognition capabilities with a variety of biomedical engineering applications.[4] The purpose of this work was to improve an existing physical model of the knee and lower limb and use Kinect to control the knee motion of the artificial limp remotely.

2. Methods

This study continued prior work, during which a physical model of the knee and lower limb was created and the knee motion was analysed. Current work was divided into three steps: First, the physical robotic knee joint was optimized and tested for a wider range of motion. Second, a computer model was created and controlled by the Kinect depth sensor. Third, the physical robotic knee and the computer model were combined, programmed, and tested for real time movement.

2.1. *Physical Robotic Knee Joint Simulation*

In previous work, a robotic lower limb was created to simulate the movement of the human knee.[5] The original knee joint included a stepper motor, an Arduino with a motor shield, two circle gears of 27 tooth each and a vertical connection with 1:1 ratio. The knee joint was connected to a construction simulating the lower leg made of light plastic material and 45 cm in length. In this work, all parts of the original construction were optimized. The following paragraphs discuss the modifications to each element of the physical model separately and their impact on the performance of the knee model.

2.1.2. Arduino

Arduino is a modern open source platform based on a microcontroller that can be reprogrammed through C language. Arduino is suitable for interactive projects as it has the ability to receive, digital or analog data. Arduino can be extended with shields. Shields are boards with microprocessors or microcontrollers that can be added on the Arduino's main board, to match the needs of a specific project. For this project, a motor shield was used that allowed to the control of the servo motor with a simple algorithm. An interface was also developed for the communication of Arduino with a PC computer and other software through a serial port.[6]

2.1.3. Servo Motor and Worm Gears

A servo motor replaced the original stepper motor of the original knee model to take advantage of its smaller size and greater power.[5] The servo motor does not accept degrees as an input since it rotates by "reading" pulses, i.e., analog signals. Consequently, the angles need to be translated from degrees to spins and from spins to an analog type of data. The analog data are then fed into the servo motor that starts spinning in the specified direction with the specified speed. The speed of the motor depends on the power applied to it, which is proportional to the distance it needs to travel (proportional control).[7] The rotation speed and the gear ratio are used to convert degrees to spins.

Several tests were performed to determine the association between degrees of rotation and time of our servo motor. The conclusion was that the motor rotated 360 degrees in 1320 milliseconds (ms) or 1.32 s. Table 1 lists the parameters of the motor servo and the gears for seven different angles that correspond to lower leg positions. These angles were selected from published work as well as physiology data for the human knee motion.[8] Note that the gear data correspond to a worm gear set that replaced a pair of wheel gears originally used in the knee model to allow for movement of larger weights. The ratio of the worm gear is 1:X, where X is the number of teeth of the second gear attached to the worm gear (driving gear). In our case X=75 so our ratio was 1:75. Our servo motor's torque was 32 oz-in (or 2.3 kg-cm) at 5V power supply and, consequently, its torque was 32x75=2400 oz-in (or 172.5 kg-cm). The speed of the servo motor was 39 rotations per minute (RPM).

2.2. *Modeling of Knee Joint*

A computer model of the knee was created with the help of Blender software. Blender is a free and open source 3D creation suite. It is a complete 3D modeling suite that supports design, rigging, animation, simulation, rendering, compositing and motion tracking, video editing and game creation.[9] Blender supports add-ons or scripts written in Python language. The designed model may be controlled by the Kinect sensor allowing for real time interaction.

2.2.1. Kinect Sensor

The Kinect (Microsoft Corp, WA, USA) sensor uses an infrared projector and a monochrome complimentary metal-oxide semiconductor (CMOS) based camera to create a depth map of an environment independent of lighting conditions. The infrared beam hits an object and its reflection is detected and processed by the CMOS to create a depth map and then infer body position allowing the recognition of movement through this process. There is also an RGB video camera recording in real time that can be used for voice recognition. Only the depth sensor was used in this study.

Journal of Physics: Conference Series **637** (2015) 012043 doi:10.1088/1742-6596/637/1/012043

Table 1: Servo motor and gear measurements used to convert degrees of rotation to motor spin. The degrees correspond to various physical leg positions or knee flexions as indicated in the first column.

Knee motion Degrees	Second Gear Number of teeth moving	Driving Gear Number of rotations	Servo Motor's Number of rotations	Time needed for servo's one full rotation (s)	Time needed for the servo to rotate (s)
67 (normal gait/level surfaces)	14	14	14	1.32	18.48
83 (stair climbing)	17	17	17		22.44
90 (sitting/rising from chairs	19	19	19		25.08
100 (advanced function)	21	21	21		27.72
106 (advanced function)	22	22	22		29.04
130 (squatting)	27	27	27		35.64

2.2.2. Blender Model, NI_Mate & Python Scripts

The creation of a model on Blender was a relatively easy process. The control of the model directly by the Kinect sensor for motion tracking was the part that posed the greatest challenge. The control was not consistently reproducible and there were often crashes in the communication between the two interfaces. To achieve a smoother communication between Kinect and Blender, the NI_Mate program was used in combination with Blender. The NI_Mate records the coordinates of the live model from the Kinect sensor and matches them to the model's coordinates created in Blender. The result is the prosthetic leg's movement according to the live model's movement. A Python script was finally implemented to calculate and display the correspondence between the computer model's knee joint degrees and the interactive model's knee joint degrees.

The block diagram of Fig. 1 presents with images the sequence of processes involved in using live knee movement to control the robotic knee.

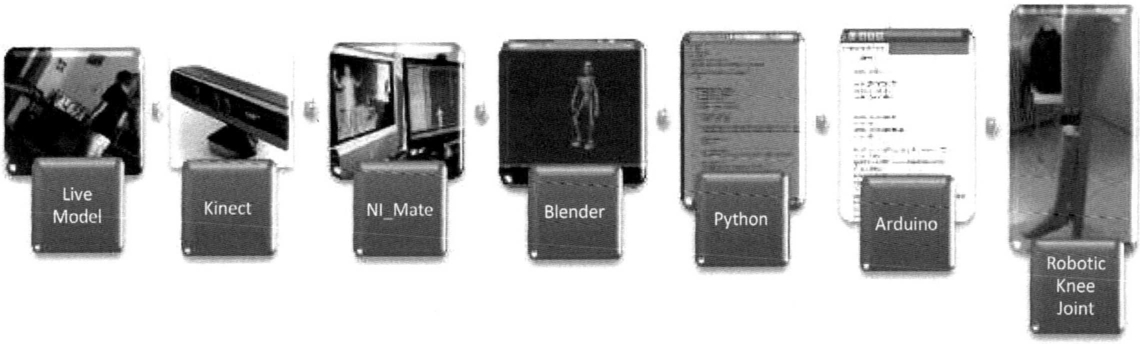

Figure 1: Block diagram of the processes involved in using a live model to control robotic knee movement.

3. Results & Conclusion

Based on the speed of the servo motor (39 RPM) and the time required for the motor to rotate for each movement (Table 1), the angular and tangential velocities for each of the six knee flexions of the simulated leg were calculated. The average angular velocity was 3.6 degrees/s with a standard deviation of 0.04. The tangential velocity was 1.6 m/s with a standard deviation of 0.02. Both results indicated a smooth and reproducible movement of the prosthetic leg independent of angle of rotation. The time required for the motion information to be transferred from the live model to the computer model and then to the prosthetic knee was negligible compared to the time required to complete the flexion in one direction. The angular velocity was the same in both directions of the movement.

In conclusion, a series of processes were implemented to achieve movement of a prosthetic lower limb from a live model. The implementation was based on Microsoft's Kinect depth camera, which, in combination with open source software communicated with a microprocessor (Arduino) embedded in a prosthetic knee to control its flexion via a stepper motor. The advantage of the current implementation is that it used low cost materials, widely available hardware, and open source software, which, however, are not always compatible. In addition to prosthetics, the proposed setup could be used for patient rehabilitation either at the hospital or at home via a remote control operation, for accurate measurements of the knee's stress and strain, for remote control of invasive procedures. The disadvantages of this project were the high use of data from each software and hardware with the result of crashes once a time.

References

1. *The Rehabilitation of People with Amputations.* World Health Organization, United States Department of Defense, MossRehab Amputee Rehabilitation Program, MossRehab Hospital, USA. 2004.
2. Ziegler-Graham K, MacKenzie EJ, Ephraim PL, Travison TG, and Brookmeyer R. Estimating the Prevalence of Limb Loss in the United States: 2005 to 2050. Archives of Physical Medicine and Rehabilitation 2008; 89(3):422-429.
3. Strbac M, Kocovic S, Markovic M, and Popovic DB. Microsoft Kinect-Based Artificial perception System for Control of Functional Electrical Simulation Assisted Grasping. BioMed Res Intern 2014; Article ID 740469, 12 pages.
4. Webster D and Celik O. Systematic review of Kinect applications in elderly care and stroke rehabilitation. J Neuroeng Rehabil 2014; 11:108-132.
5. Velmahos I and Zekio M. *Robotic knee model.* Department of Biomedical Engineering, Technological Educational Institute of Athens. May 2014.
6. CEO, "Arduino," Arduino Organization, [Online]. Available: http://arduino.cc/en/guide/introduction.
7. http://mechatronics.mech.northwestern.edu/design_ref/actuators/servo_motor_intro.html
8. Chakravarthy YK, Kumar RH, Deepika B, and Srinath A. Haptics based robotic prosthetic leg. Advances in Robotics, Mechatronics, and Circuits. ISBN: 978-1-61804-242-2.
9. Blender Foundation, "Blender," Blender Organization, [Online]. Available: http://www.blender.org/features/.

Institute of Physics
Dirac House, Temple Back
Bristol BS1 6BE UK

ISSN: 1742-6588
ISBN 978-1-5108-1315-1

9 781510 813151